Loon Lessons

Loon Lessons

Uncommon Encounters with the Great Northern Diver

James D. Paruk

University of Minnesota Press
Minneapolis
London

MIN NE SO TA

The University of Minnesota Press gratefully acknowledges the generous assistance provided for the publication of this book by the Hamilton P. Traub University Press Fund.

Published by the University of Minnesota Press
111 Third Avenue South, Suite 290
Minneapolis, MN 55401-2520
http://www.upress.umn.edu

ISBN 978-1-5179-0940-6 (hc)
Library of Congress record available at https://lccn.loc.gov/2020053628.

Printed in Canada on acid-free paper

28 27 26 25 24 23 22 21 10 9 8 7 6 5 4 3 2 1

To Olivia and Emily, my passionate and miraculous daughters
who provide me with daily inspiration, a gift I cherish

Contents

Preface ix

1. **In Search of the Ancestral Loon** 1

2. **Selected by Nature**
A Tale of Two Birds 17

3. **What a Drag!**
The Inner Workings of a Master Diver 31

4. **Pairing Up**
The Behavioral Ecology of Loon Courtship 49

5. **Taking Turns**
Nesting Behavior and Ecological Trade-Offs 67

6. **Wails, Yodels, and Tremolos**
The Call of the Loon 81

7. **A Vigilant Bird**
Parents, Chicks, and the Breeding Loon Family 97

8. **More Than a Foot Waggle**
The Fascinating World of Loon Behavior 113

9. **Loons on the Move**
The Strategy and Dynamics of Migration 127

10. Not Your Typical Snowbird
The Loon's Winter Ecology 145

11. Saving the Loons We Love
Conservation Threats 167

12. Loon Watch
Adapting to a Changing World 189

Epilogue 211

Acknowledgments 213

Resources for Loon Conservation 217

Index 219

Preface

To appreciate any organism, I am convinced we do not need years of training—all we need to do is watch our children marvel at a deer or a squirrel from a window. A sense of wonder and our innate curiosity can lay the foundation for developing and maintaining an appreciation for the natural world. Most of us vividly remember our first experience with a loon and that sense of wonder we felt when we heard its call or watched it swim near us. I have great appreciation for the Common Loon and have been fortunate as a scientist to study them for three decades. If you have had the pleasure to see or hear them, they need no introduction. They are majestic, regal looking, with never a feather out of place, and their mysterious and eerie vocalizations are unforgettable. Loons permeate our culture. There are loon festivals across North America, filled with informative talks and loon-calling contests. They adorn mailboxes, road signs, mugs, pillows, coasters, calendars, key chains, and trinkets, which litter our homes and cabins. Loons have a large following indeed, and I, too, am not immune to their seduction.

But what can explain this following? Why do we love loons so much? What is it about them that captures our imaginations? I think, in part, we are drawn to loons because they are stunning to look at and captivating to listen to. Their striking black-and-white markings are unique among birds, and their calls are mysterious and haunting, unlike anything else. Loon vocalizations evoke such strong emotions in us that their calls are even dubbed into some Hollywood films. John James Audubon captured their majesty and eloquence when he wrote of them in his book *Birds of America* (1827–38) nearly two hundred years ago:

> The Loon, as this interesting species of Diver is generally called in the United States, is a strong, active, and vigilant bird. When it has acquired its perfect plumage, which is not altered in colour at any successive moult, it is really a beautiful creature; and the student of Nature who has opportunities of observing its habits, cannot fail to derive much pleasure from watching it as it pursues its avocations. View it as it buoyantly swims over

> the heaving billows of the Atlantic, or as it glides along deeply immersed, when apprehensive of danger, on the placid lake, on the grassy islet of which its nest is placed; calculate, if you can, the speed of its flight, as it shoots across the sky; mark the many plunges it performs in quest of its finny food, or in eluding its enemies; list to the loud and plaintive notes which it issues, either to announce its safety to its mate, or to invite some traveller of its race to alight, and find repose and food; follow the anxious and careful mother-bird, as she leads about her precious charge; and you will not count your labour lost, for you will have watched the ways of one of the wondrous creations of unlimited Power and unerring Wisdom. You will find pleasure too in admiring the glossy tints of its head and neck, and the singular regularity of the unnumbered spots by which its dusky back and wings are checkered.

Audubon's knowledge of loon natural history is impressive, especially in an era without access to modern optics, transportation, or advanced field guides. He got many aspects of their breeding, wintering, and migration biology right, using a keen and well-traveled eye. He noticed, for example, that during the breeding season loons initiate breeding in the southern part of their range one month sooner than in the northern part, that they are more numerous in Canada than in the United States (still true today, if not more so), and that females are smaller than males. During the winter he further observed that loons occupy coastal waters along the entire Eastern Seaboard, from Labrador to Florida, that they molt their feathers, and that besides eating fish, they will consume invertebrates like crabs on occasion. He noticed that during spring migration loons will stop along major rivers because they are a bountiful source of fish, and during fall migration may fly 1,000 to 2,000 feet or more above the land. He refuted the notion that loons become harbingers of foul weather by increasing their frequency of calling because he saw no direct evidence to support that. Despite a few inaccuracies, which modern readers can easily overlook, Audubon's writing about loons exudes the excitement and wonder that we all feel when we catch a glimpse of these magnificent birds.

I recall vividly the moment when the seed of passion and wonder for loons began in me: it was when I first touched one in the wild. In late July 1989, I was at Seney National Wildlife Refuge in Michigan's eastern Upper Peninsula to help a college buddy, Dave Evers, catch loons as part of his master's research at Western Michigan University. How do you catch a loon? You go out at night in a boat, approach a family of adults with chicks, shine a bright light to disorient them, and, with alacrity, scoop them up before they dive. One night, Dave adroitly lifted an adult loon out of the water and placed it on my lap, tucking its bill under my armpit. Was I nervous? Absolutely. The loon struggled, and I used all my strength to keep it under control, to prevent the wings from opening. I felt the warm liquid of excrement roll down my non-waterproof pants.

After a bit of a struggle, the loon relaxed, and Dave placed a metal band with a nine-digit number around its leg. The band could later be used to identify this individual should it be recaptured or wash up on shore in Lake Michigan or South Carolina. Then he placed colored bands around its legs so (in theory at least) one could identify the bird at a glance without viewing the tiny nine-digit number. He carefully took feather and blood samples from the loon. With my fingers, I noticed how stiff the feathers were along its back and wings and felt its heartbeat through my hand. I was so focused on this moment that I was unaware of the passing of time. Dave asked me to reorient the loon so that it was sitting on my lap with its bill pointing away from me. The loon wailed, and after a few pictures were taken, Dave took the bird from me and released it in the water. That single experience ignited my passion for loons and changed the course of my life.

After helping Dave that summer, I read Judith McIntyre's book *The Common Loon: Spirit of Northern Lakes,* published the year before by the University of Minnesota Press, and I was impressed by her knowledge of loons. McIntyre had been studying them for nearly two decades and was a treasure trove of insight. She identified a number of questions for future researchers: How long do pairs stay together? Do the sexes play an equal role in incubation and chick provisioning? Do loons mate for life? Armed with Dave's groundbreaking capture technique, I enrolled in a doctoral program and investigated sex roles between loons during different stages of the breeding season (pre-nesting, nesting, and post-nesting). I was fascinated by loons and kept plugging along studying them, at different intensities, for the next twenty-seven years. My research has taken me all over North America, from southern California to Alaska and from Washington to Maine. I have captured and handled more than two hundred loons, spent more than five thousand hours quantifying their behavior on both freshwater lakes and the ocean, and authored or coauthored more than two dozen scientific articles. *I know loons.* And I say this, of course, with the caveat that, like any scientist, I still have much to learn.

At my core I am a teacher and a science educator: that is why I wrote this book. I am not exactly Bill Nye (not a fan of ties), but my zest for educating others about loons is in my DNA. I cannot help myself—just ask my daughters as they roll their eyes (again) when they see me introduce myself to a stranger on a plane and start talking about loons. My college professors showed me the wisdom in asking questions to get beyond the obvious, and I have taken their modeling to heart. When I see a loon on the water, my mind starts racing, and I begin asking questions. Why does a loon have a red eye? Why does it have a necklace or white spots on its back? Why does one loon lower its head and bill as it approaches another loon? How do loons communicate with each other? Getting the facts right helps us make sense of an animal's behavior. For example, some loons will sneak furtively into another loon's territory and, if given the opportunity, will attempt to drown and kill the chicks. Why would a loon do that? To understand and interpret a behavior like infanticide requires a framework for observations. Animals hardly play nice. The framework that unites all of biology is evolution.

Darwin got it right. Animals are shaped and act in ways that maximize their survival and ability to reproduce. Camels have long eyelashes, snowshoe hares have large feet, and walking sticks look like twigs. Over time, these modifications and adaptations were incorporated into the DNA of these animals. Adaptations allow individuals with them to survive or reproduce better than individuals without them. Loons are shaped and behave in certain ways because generations before benefited from these DNA modifications that improved their chances of success and reproduction. I have had training and experiences that allow me to see organisms in a certain light. Not better, just different. With these eyes I wish to convey to you the story of the Common Loon and the lessons we can learn from it.

Every aspect of loon anatomy, behavior, and life history is rooted in evolution: it can be no other way. Many birds can swim on the surface of water, but few are able to dive. For a dive of any significant duration, a bird must move through the water under its own power, overcoming incredible drag forces. The fact that proportionally so few birds can do this attests to the difficulty of this feat. Loons, along with other diving birds, exhibit a number of specializations in both their skeletal and muscular anatomy, which are introduced in the first few chapters of this book. I explore some little-known physiological features, but most of these pages are devoted to loon behavior, including fascinating research on loon communication, for that is what we see and hear when we encounter loons. The later chapters examine migration and winter ecology, a subject I have researched for ten years. Finally, *Loon Lessons* introduces loon conservation and identifies some of the threats loons face in an increasingly crowded world.

It has been a long journey. I have swatted many blackflies and mosquitoes, perspired under the hot sun, shivered at midnight while catching loons in the middle of winter, stayed up until the early hours of dawn, and traveled the North American continent, all in the hope of understanding this bird. How much more insight would I have gained if I could only have been inside the head of a loon for a day? A month? A year? Alas, it was not to be. Therefore, with the benefit of learning from many other loon researchers who have collectively increased our knowledge and understanding of this transformative bird, I express my thoughts as a curious naturalist. At the many talks I have given there is no shortage of questions, stories, and curiosity from those in attendance. I appreciate all they share, but I especially am grateful for their passion. Loons really do have an undeniable following. Audubon would approve.

1

In Search of the Ancestral Loon

In June 2011, I was lying flat on the ground, cold, wet, and surrounded by ravenous mosquitoes. I was chilled. To get warm, I could have stood up and moved around, but I had a job to do, and for the time being it required I wait patiently, stay out of sight, and watch for a loon to swim back to its nest. Just the day before, I had arrived in Barrow, Alaska, and this morning I had taken a thirty-minute helicopter ride over tundra, dotted with numerous small lakes and crisscrossed trails. After a few more minutes in the air, we had looked down and spotted migrating caribou. Having grown up watching Mutual of Omaha's *Wild Kingdom* while dreaming of wilderness adventures like this, I almost pinched myself.

Our camp was roughly 60 miles south of the Beaufort Sea and 200 miles north of the fabled Brooks Range. After introductions, I stepped over the electric fence that encircled our tents to deter grizzly bears from entering. I was one of several biologists at the camp. Most of us were involved in a project investigating the Yellow-billed Loon, the rarest of the five loon species. After my arrival, I went out with Joel Schmutz, senior scientist at the U.S. Geological Survey and the project leader, to assist him in catching Greater White-fronted Geese. I was excited to walk on the tundra, and along the way we spotted a pair of Yellow-billed Loons, close to the camp. Again, I needed to pinch myself. Before long, we would also see Red-throated and Pacific Loons. Later that day I met Ken Wright, an independent Canadian biologist who had been working with Joel for the past few years. His enthusiasm for the tundra and the project was contagious. Around nine o'clock that evening, I joined Ken and two other biologists to lend a hand in catching the Yellow-billed Loons near the camp.

As we approached a nest, one of the pair got off it and then proceeded to call and swim with its mate to the other side of the lake. We took the eggs, placed them securely inside a lined Tupperware container, and replaced them with wooden models. This way the eggs would be safe and not exposed to aerial predators, like Glaucous Gulls, while we waited for the loons to get back on the nest. We entered the water in front of the nest and floated out a net, securely tying it to the bottom. In theory, the net would be hard to detect, and the adults' motivation to get back on the nest would override their sense of caution about this new object in their environment. The other biologists remained near the nest, lying flat and out of sight, while I walked roughly a hundred yards away, before I too fell to the ground.

An hour elapsed, and the cold, wet tundra and mosquitoes were testing my patience, when suddenly one of the loons began making its way back to the nest. My job was to notify the other biologists when a loon was within range of the net, but with my face on the ground, it was hard to make that determination. Not yet, I said to myself. Wait until it is right next to its nest; then it should be over the net. I waited a few more minutes. It was our decisive moment. It would either dive into the net and get caught, or swim over it. "*Now,*" I spoke into the radio. They jumped up and shouted, and the loon dove to escape them, right into the net. By the time I arrived at the capture site the other biologists had the loon out of the net. We walked with it to shore, sat down, and began data collection, measuring its leg and bill, and placing colored bands on its legs. We also drew blood and collected a couple of feathers, which would later be used to assess contaminant levels, such as mercury. I noted how remarkably calm the bird was during this whole process, unlike most Common Loons I have handled.

As this was my first capture of a Yellow-billed Loon, the other biologists deferred to me to release it. I appreciated the moment. I picked it up and immediately noticed how large it was compared to loons I had caught in the Midwest. I had a hard time getting my small hands and short arms around the body, but I managed, in part because the loon remained calm. I stared at its yellow bill and wondered if the difference between it and the black bill of the Common Loon was a single mutation. I looked deep into its red eyes, and the bird matched my stare. My graduate student Brandon Braden took a few pictures of me with the loon (I did not have to feign smiling). Finally, it was time. I walked into the water, lowered my frame, and gently nudged the bird to swim away from me. The loon cooperated. We gathered all our gear and replaced the eggs before walking to the next lake to catch more loons. I never thought I would be this fortunate. Where I grew up in Hamtramck (an enclave of Detroit), wilderness and encounters with wildlife resided strictly in my imagination. To this day, my first encounter up close and personal with a Yellow-billed Loon ranks as one of the highlights in my life as a biologist.

What Makes a Loon a Loon?

•

Let us start with some basics. What exactly is a loon and what makes them different from other similar-looking waterbirds? A loon is a large, piscivorous (fish-eating) diving bird that propels itself underwater by large webbed feet, rather than using its wings like a penguin does. Loons breed on freshwater lakes across North America, and winter in predominately marine environments (i.e., ocean, coastal bays, and coves). Loons rarely get out of the water. Their legs are so specialized for diving and positioned so far aft that standing, let alone walking, is nearly impossible; hence, the adoption of the Scandinavian word for "clumsy," *lom* or *lumme,* which stuck as their common name. A loon's breeding plumage is distinctive, with the neck having either short or long vertical black and white stripes. Their bills are long, thick, and straight. They have narrow, pointed wings, consisting of ten primary feathers (outer wing) and twenty-two or twenty-three secondary feathers (inner wing). Loons have three fully webbed front toes, and the fourth toe, the hallux, is long and elevated above the others. They have fourteen to fifteen cervical (neck) vertebrae. Their unique vocalizations include a yodel and a wail.

Other diving birds, such as freshwater-nesting grebes and mergansers and marine-nesting auks, guillemots, murrelets, auklets, and puffins, do not share these same physical traits. Certain molecular differences (DNA, mitochondria, proteins) also separate loons from other diving birds. Because of their many unique features and traits, loons are placed in their own order (*Gaviiformes,* L, GAV-ee-i-form-eez), family (Gaviidae) and genus (*Gavia*). Because loons are so different from other diving birds, several questions about their origins surface: Who among the diving birds are the loon's nearest relatives, and when did they last share a common ancestor? How far back in the fossil record can we find loons or loon-like birds? When did the birds that we now know as loons appear?

In Search of the Ancestral Loon

To answer these questions, we have to go back 150 years to the summer of 1871, in Kansas, with not a lake in sight. Ten students from Yale University were busy assisting their vertebrate biology professor, Othniel Charles Marsh, in finding fossils. Marsh was one of two leading paleontologists of his time who collectively increased the number of dinosaur species from 9 to nearly 150. He chose this part of Kansas because the bedrock was made of ancient marine limestone from the Western Interior Seaway, essentially a mini-ocean that sat over what are now the prairie states. It was tedious, challenging work, and the expedition would find several interesting fossils, including *Stegosaurus, Allosaurus,* and *Triceratops,* but they also found an extraordinary one that was just under 5 feet long and headless. Marsh correctly deduced that the animal was not a small dinosaur but a bird. However, it was a diving bird that ate fish—unlike any bird previously discovered. He named the fossil *Hesperornis regalis,* which means "regal western bird."

Even though the limestone where *Hesperornis* was collected was 75 to 85 million years old, it could not have been the first diving bird. We know this because its skeletal system was already greatly modified and specialized for diving. *Hesperornis* was no primitive diving bird nor a transitional fossil. Its feet were positioned at the rear of its body, and its forelimbs were greatly reduced (thus, it could not fly). Its leg bones were flattened, resembling a canoe paddle, allowing maximal thrust on the backstroke, and when twisted, minimal resistance on the return stroke. The origin of the first diving birds had to predate *Hesperornis* by at least 10 or 20 million years, which pushes the evolutionary history of diving birds back to approximately 100 million years ago (the first terrestrial birds appear in the Jurassic, 150 million years ago). Since *Hesperornis,* paleontologists have unearthed at least thirteen more diving bird species.

A quick look at the anatomy between the two species reveals that loons cannot be descendants of *Hesperornis.* Why? For one big reason: the numbers of bones in the forelimb (the wing, or arm) do not match up. *Hesperornis* has only one arm bone in its forelimb, the humerus, while loons have three (humerus, radius, and ulna). It is highly improbable that loons grew a radius and ulna after *Hesperornis* lost the ability to generate them. Furthermore, *Hesperornis* had a unique bone in the roof of the mouth that no modern bird possesses. It was very large, up to 6 feet long, and an accomplished diver; however, it left no avian descendants, at least none that shows up in today's modern birds.

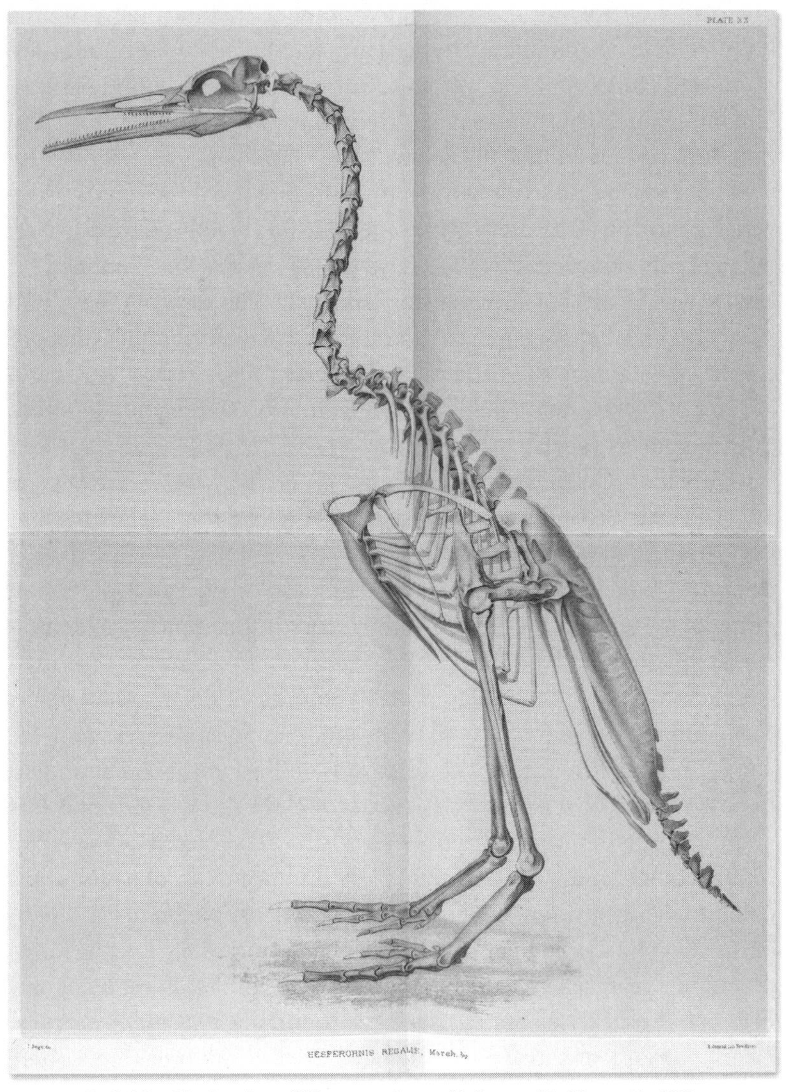

O. C. Marsh's drawing of *Hesperornis regalis,* located in Kansas in 1871. Courtesy of the Biodiversity Heritage Library.

The most recent molecular data comparisons among birds reveal that the nearest relative of loons is penguins (and possibly petrels). It has been a long road to reach this conclusion, and for over a century, many groups, such as hesperornithine birds, grebes, petrels, gulls, penguins (among a few others), have been proposed to be loons' nearest relatives. The confusion was due to convergence, a phenomenon where organisms occupying the same environment (habitat) look similar, because of local adaptation, but are unrelated. Cacti, for example, are unique to North America, yet go to Africa and you will find similar-looking plants, with the waxy cuticle and greatly reduced leaves. But that does not mean they are related. They have similar traits because they adapted to essentially the same environment, a desert, and to save water, they evolved a water-saving outer skin and greatly reduced leaves. Diving birds share similar anatomical traits (webbed feet, flattened leg bones, legs far aft) because they too occupy similar environments, and it becomes challenging then to sort out how many of the similarities are due to convergence or to shared ancestry. On the surface, grebes and cormorants share some similar traits with loons, but again, that does not mean they are related.

Early loon-like and penguin-like fossils both appear in the Eocene epoch, 33.9 to 56.0 million years ago. Penguin fossils appear a little earlier (about 50 to 55 million years ago) than loons (40 to 50 million years ago). The main problem with interpreting the origin and evolution of either of these groups is that the fossil record is incomplete (paleontologists will often find just a single leg or arm bone), and what does exist is discontinuous, both geographically and geologically. This means that certain types of fossils tend to be abundant in some areas and uncommon to rare in others. They also tend to be common to abundant in some strata (a recognizable layer of rock in the ground) and less common in other strata. There is always some subjectivity based on interpretation and experience, and the interpretation of a story may change slightly given new information. It appears that loons and penguins shared a common ancestor at the beginning of the Eocene (55 to 56 million years ago), which makes them one of the oldest living lineages of birds. Old leg bones have been found in Chile and Antarctica, evidence that loons may have originated in the Southern Hemisphere and then dispersed to the Northern Hemisphere.

Early loon-like fossils reveal they were not as specialized for diving; the positioning of the legs was not as far back on the body, and the leg bones were not as flattened. Several authorities feel the direct descendants of loons may trace back to an extinct group of birds in the genus *Colymboides.* From fossils, we know that *Colymboides* lived in Europe 20 to 37 million years ago, was smaller than the modern-day loon (closer in size to a small grebe or teal), and was not as specialized for diving. Yet, the two genera (*Gavia* and *Colymboides*) share many similarities, which suggests a link exists between them. So, *Colymboides* could be the ancestor of *Gavia,* or it could not. Approximately 15 million years ago, the earliest known member of the genus *Gavia,* from a fossil in Czechoslovakia, appears. This fossil came from the same strata as *Colymboides.* Did *Gavia* split from *Colymboides* prior to 15 million years ago? Perhaps. For the next 10 million years or so, few other fossil loons appear in the fossil record. Then during the Pliocene epoch (2.6 to 5.3 million years ago), several different-sized skeletons of *Gavia* (*G. concinna, G. howardae, G. palaeodytes,* and *G. portisi*) show up in deposits from California, Florida, and Italy, and shortly afterwards, in the Pleistocene (fewer than 2.6 million years ago), the skeletons of all of our current loon species show up.

Loon Diversity

There are five species of loons worldwide: Red-throated (*G. stellata*), Pacific (*G. pacifica*), Arctic (*G. arctica*), Common (*G. immer*), and Yellow-billed (*G. adamsii*), all likely descendants of loon species that lived during the Pliocene. They reside in the Northern Hemisphere, in North America and Eurasia. The most notable differences among the five species are body size, bill size and shape, and plumage. Based on body size, loons are split into three groups: small, medium, and large. The smallest, the Red-throated, weighs between 3 and 4.5 pounds. The Pacific and Arctic Loons are both medium sized, 3.5 to 4.5 pounds and 4.5 to 7 pounds, respectively. The Common Loon and the Yellow-billed Loon are the two largest loon species, each weighing more than 7 pounds. Over nearly all of its range, the Common Loon is smaller (7 to 12 pounds) than the Yellow-billed Loon (12 to 16 pounds), with the exception of New England, where a few male Common Loons tip the scales at 16 pounds. As loon species increase in body size, so does their bill size. Bill shape varies among the group. In the Red-throated and Yellow-billed it is slightly upturned, but it is straight in the Pacific, Arctic, and Common.

As for breeding plumage, loon spotters take note! The Red-throated Loon lacks the distinctive black-and-white checkered markings present in the other four species. Its head is gray, and the ventral side of the neck is red. Both Pacific and Arctic Loons are easily separated from the other species but are difficult to tell apart from one another. They each have black-and-white markings on the back, gray heads, and distinctive black on the ventral side of the neck. We can use size to separate the larger Arctic Loon from the smaller Pacific Loon, but that feature may be difficult to use in the field, especially at a great distance. Instead, we can see that the back of the neck is darker, and the white vertical lines more pronounced on the Arctic compared to the Pacific. However, the best trait to distinguish between the two species is the white flanks present near the tail of the Arctic, which are black in the Pacific. Common and Yellow-billed Loons are easy to separate because of the noticeably paler bill in the latter.

The Common Loon has the southernmost breeding distribution of all loon species. It is present in the border states of the United States and Canada and nests on lakes bordered with trees. The other four species nest on lakes in the Arctic tundra, generally devoid of trees. Globally, the Red-throated is the most widely distributed of all the loons, ranging across the upper parts of North America and Eurasia. One can argue that the Red-throated is unique among the loon species for a number of reasons. First, it is the most marine of the loon species, regularly making forays to feed in the ocean during the breeding season. They nest on small ponds close to the ocean and leave them frequently to make flights to open water to bring back food for their young. Second, they can gain flight without the long runway (about 15 yards) that is more necessary for the other loon species. Third, both female and male Red-throated Loons yodel (a type of vocalization), whereas in the other four species, only the males yodel. Fourth, rather than molt their flight feathers in winter, like the other species, they molt them in the fall, at a staging ground, like the Gulf of St. Lawrence or the Great Lakes. Lastly, they are the most gregarious of all the loon species, particularly so in the winter when they form immense flocks of one thousand individuals or more. These flocks forage cooperatively,

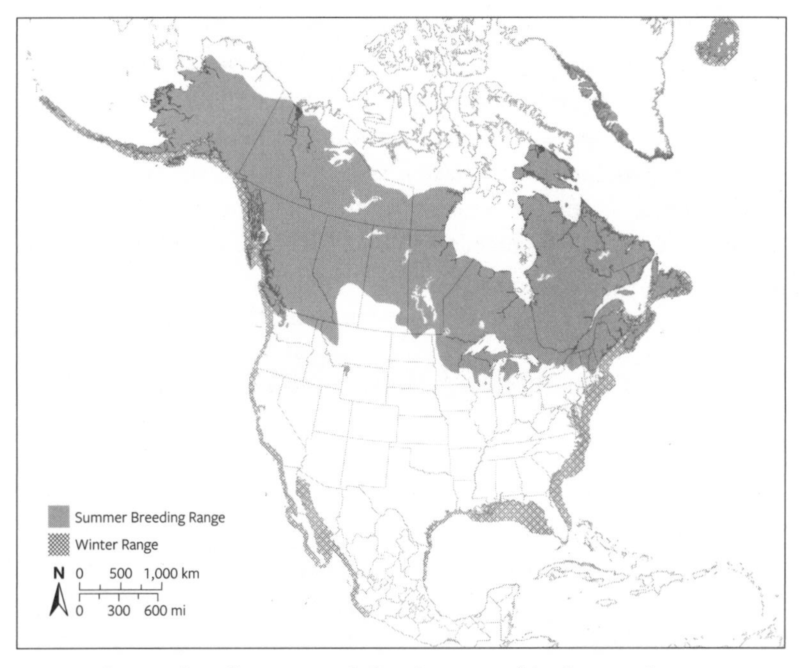

Summer breeding range and wintering range of the Common Loon.
Map by Mark Burton, Biodiversity Research Institute.

driving fish into shallower water. Because of their flocking foraging behavior, many drown in commercial fishnets. If I had another lifetime, I would study Red-throated Loons.

The Pacific Loon breeds in northern Canada beginning on the western edge of Hudson Bay, west to Alaska, and across the Bering Sea to the easternmost parts of northern Russia. The range of the similar-looking Arctic Loon overlaps with the Pacific Loon in far eastern Russia but then extends across northern Eurasia as far as Norway. Arctic Loons breed in only one place in North America, the Seward Peninsula of Alaska. With regard to visual appeal, my vote is for the Pacific as the stateliest of all the loon species. Because they occupy the same niche (more on that in a moment), the differences in the ecology and breeding biology between the Pacific and Arctic species are not great, so I will treat them more or less as the same. To reduce competition with the Red-throated Loon, they occupy larger lakes and ponds, but not the largest, because those tundra lakes are occupied by Yellow-billed Loons. Therefore, in a sense, Pacific and Arctic Loons face some constraints. Like all loons, during the breeding season they are intolerant of loons of the same species (intraspecific aggression) and of other species (interspecific aggression). They prefer nesting on islands if available. They lay two eggs, incubate them for just under a month, and are good providers for their chicks, just like the other loon species. Pacific Loons winter predominately along the western North American coast, traveling as far south as the Baja peninsula and the Gulf of California, whereas the great majority of Arctic Loons winter along the Asian coast south to the Baltic Sea. Like other loons, they are vulnerable to oil spills during migration and winter.

The Common and Yellow-billed are large-sized loons. They breed in different locales, which minimizes competition between them. The Yellow-billed breeds on tundra lakes in the far north, whereas Common Loons nest farther south, in lakes surrounded by northern temperate or boreal forests. The Yellow-billed Loon's breeding range is restricted to north of the Brooks Range in Alaska and extends both west and east from there, always in the high Arctic. To the west, its range traverses the Bering Sea and most of Eurasia, and to the east, makes its way across Canada as far as Hudson Bay. The Common Loon has the broadest geographic range of any loon in North America. It is the only loon species to breed in the contiguous United States, across nearly all of Canada, and in most of Alaska. The Common Loon also nests in Greenland, Iceland, and Great Britain. In North America, the Yellow-billed has the most restricted geographic range of all the loon species and is the rarest. In Alaska, there are roughly a thousand pairs. Because of its rarity and restricted breeding range, the Yellow-billed is a favorite among North American birders and one of the most challenging to check off one's list.

Loons as a group have very similar DNA. We know this because some members of this group hybridize with other species. But how do we account for the different species of modern loons? When did they evolve? Molecular data supports the notion that the Red-throated is the oldest of the loon species and likely split from the others 3 to 4 million years ago. This event was followed by another (possibly a glacial episode) that saw Pacific and Arctic Loons charting their own evolutionary trajectory, apart from Common Loons. Then another event, in this case, a definite glacial episode, roughly a million years ago separated Common and Yellow-billed Loons.

Since the Red-throated is more marine than the other species and nests close to coastal waters, it is thought that some Red-throated individuals moved more inland to breed and became geographically isolated from the original population. This new incipient population would have reduced competition with the parent population. Around 2 to 3 million years ago, this newly formed population of loons split again, giving rise to Pacific and Arctic populations and Common Loon populations. The Pacific and Arctic diverged geographically east and west, while the Common diverged and moved farther south. Then, around 1 million years ago, the Yellow-billed Loon split from the Common Loon. The thinking is that a glacial event split the parent Common Loon population into two, one that remained primarily north of the glacier (in refugia), and the other migrating westward and southward off the Asian coast. The evidence for this is that some Yellow-billed Loons breed in Alaska and the Northwest Territories and take a longer route, migrating west across the Bering Sea and southward to the Asian coast, instead of flying south and wintering off North America.

The Loon's Ecological Niche

•

Loons differ from other similar diving birds in a number of physical traits, plumages, and vocalizations. They are unique, or monophyletic, in that they have been on their own evolutionary trajectory for more than 50 million years; they have no real close avian group (penguins being the closest). Yet, there are many different groups of diving birds (grebes, mergansers, guillemots, puffins) that can compete with them. So how do loons minimize competition with these groups of birds? And how do they minimize competition with each other? To address these questions further, let's consider an ecological concept, the niche.

The concept of the niche helps us better understand how the great diversity of life forms share our planet. It is the role or position taken by an organism in its community. Animals that make up a community can be divided by their diet or food preference (called a guild). For example, wolves (carnivores) eat meat, deer (herbivores) eat plants, bats (insectivores) eat insects, and so on. Within each guild, animals can be divided by body size. Canid carnivores sort out nicely by body size—wolf, coyote, and fox—as do members of the deer family—moose, elk, and deer. Guilds can also be sorted out by time of day when an animal is active (nocturnal, twilight, diurnal). For example, swallows, swifts, and bats feed on insects, but they minimize competition by feeding at different times of the day. Swallows fly during the day, swifts mostly during twilight hours, and bats at night. Lastly, similar species of animals sort out by occupying different habitats (the polar bear prefers the dry, cold conditions of the Arctic, and the sun bear prefers the humid rain forests of the tropics), or within a given habitat by vertical space (there are up to five different canopy layers animals can occupy). Ecologists noticed a pattern; no two species of animals can occupy the same niche, as there would be too much competition for limited resources. Over time, each animal carves out its niche. What is the loon's niche, and how does it differ from that of other waterbirds?

Remember, a loon is an obligate waterbird, a piscivore, that spends nearly all its waking hours in the water. It breeds on freshwater lakes and winters in marine coastal environments. Today, there are numerous piscivorous species of birds, divided into guilds based on their breeding habitat: marine or freshwater. The marine guild spends time in the ocean and breeds in colonies on islands (e.g., puffins, murres, murrelets); the freshwater guild inhabits lakes and ponds and generally does not breed in colonies (e.g., most grebes, mergansers, and loons; gulls are an exception). Therefore, loons are separated spatially from marine diving birds during the breeding season but overlap with the piscivorous, freshwater breeding guild. But how is niche overlap minimized within this guild?

I sorted members of the dog and deer family by body size and will do the same for waterbirds, using grams as a proxy for body size (454 grams = 1 pound). There are six grebe and three merganser species that breed in North America, and they range in body size from 400 to 1,700 grams, and 750 to 1,700 grams, respectively. Collectively, these two groups of birds can be considered small-sized divers (less than 1,700 grams). An example of a medium-sized diving bird is the Double-crested Cormorant (1,700 to 2,700 grams), and of a large-sized, the Common Loon (more than 3,400 grams). By virtue of their size relative to other freshwater guild species, the loons are like the moose of the deer family and occupy a distinct niche from these other birds. Because of their larger size and associated larger bill, loons are able to dive deeper, stay underwater longer, eat larger fish, eat different fish species, and crush larger crustaceans than their counterparts.

Studies of animals have found that body size is related to dominance, meaning larger individuals tend to win disputes over smaller individuals. The larger male Herculean beetle, the hippopotamus, and the bighorn sheep, for example, dominate over their smaller-sized rivals. Loons are big, and this size advantage allows them to dominate other birds using the same lake habitat, effectively reducing competition and widening the gap between the niches even more. We can see this in part by looking at where the other freshwater diving birds nest. Grebes generally nest on smaller waterbodies than loons, and if you find both species nesting on a lake, the grebe will nest away from the loon to reduce conflict and competition. Mergansers typically breed more in the backwater of a lake or river and nest in tree cavities, further widening the space between the two niches. Four of the five cormorant species in North America breed along the coast, and only one cormorant, the Double-crested, breeds both along the coast and on interior islands, typically in large freshwater lakes (including the Great Lakes). So, when additional resources are needed to feed their young, both the Double-crested Cormorant and Common Loon are spatially separated. In the winter, both of these species may occupy similar coastal environments, but subtle differences in diet likely reduce niche overlap. The cormorant feeds on smaller or different fish because it has a more flexible neck (due to more cervical vertebrae) than the Common Loon.

Loons minimize niche overlap with each other by varying in body size, habitat type, geography, nesting lake size, food preference, and so on. There are only so many ways ponds and lakes in the Arctic tundra can be separated, and normally three is the highest number of loon species occupying the tundra at any one time (Red-throated, Pacific, and Yellow-billed). I was fortunate to collaborate with Diana Solovyeva, a Russian ornithologist, from 2011 until 2015 as she and her staff observed four species of loons nesting in relative proximity on the Kyttyk peninsula outside Chukotka in far eastern Russia. Because of its isolated geography, this was a unique situation. Given their niche, geographical distribution, and biological constraints, five different loon species fill the available niche space in the Northern Hemisphere, but only one, the Common Loon, breeds on lakes in the Lower 48 and southern Canada, where people predominately inhabit. This is why Common Loons are so popular—because we see and hear them all summer long.

Further Reading

Bell, A., and L. M. Chiappe. 2016. A species-level phylogeny of the Cretaceous Hesperornithiformes (Aves: Ornithuromorpha): Implications for body size evolution amongst the earliest diving birds. *Journal of Systematic Paleontology* 14 (3): 239–51. https://doi.org/10.1080/14772019.2015.1036141.

Boertmann, D. 1990. Phylogeny of the divers, family Gaviidae (Aves). *Steenstrupia* 16 (3): 21–36.

Carroll, S. B. 2005. *Endless Forms Most Beautiful.* New York: W. W. Norton and Company.

Chase, J. M., and M. A. Leibold. 2003. *Ecological Niches: Linking Classical and Contemporary Approaches.* Chicago: University of Chicago Press.

Colbert, E. H. 1984. *The Great Dinosaur Hunters and Their Discoveries.* New York: Dover Publications.

Davis, L. S., and M. Renner. 2003. *Penguins.* New Haven, Conn.: Yale University Press.

Dawkins, R. 1996. *Climbing Mount Improbable.* New York: W. W. Norton and Company.

Gould, S. J., and E. S. Vrba. 1982. Exaptation—A missing term in the science of form. *Paleobiology* 8: 4–15.

Hackett, S. J., R. T. Kimball, S. Reddy, R. C. K. Bowie, E. L. Braun, M. J. Braun, J. L. Chojnowski, et al. 2008. A phylogenomic study of birds reveals their evolutionary history. *Science* 320: 1763–68.

Hardin, G. 1960. The competitive exclusion principle. *Science* 131: 1292–97.

Haynes, T. B., J. A. Schmutz, M. S. Lindberg, K. G. Wright, B. D. Uher-Koch, and A. E. Rosenberger. 2014. Occupancy of Yellow-billed and Pacific Loons: Evidence for interspecific competition and habitat mediated co-occurrence. *Journal of Avian Biology* 45: 1–9.

Hutchinson, G. E. 1957. Concluding remarks. *Cold Spring Harbor Symposium on Quantitative Biology* 22 (2): 415–27.

Jetz, W. G., H. Thomas, J. B. Joy, K. Hartmann, and A. O. Mooers. 2012. The global diversity of birds in space and time. *Nature* 491: 444–48.

Johnsguard, P. A. 1987. *Diving Birds of North America.* Lincoln: University of Nebraska Press.

Lindsay, A. R. 2002. Molecular and vocal evolution in loons (Aves: Gaviiformes). Ph.D. diss., University of Michigan, Ann Arbor.

McIntyre, J. W. 1988. *The Common Loon: Spirit of Northern Lakes.* Minneapolis: University of Minnesota Press.

McIntyre, J. W. 1994. Loons in freshwater lakes. *Hydrobiologia.* 279/280: 393–413.

Olson, S. L., and P. C. Rasmussen. 2001. Miocene and Pliocene birds from the Lee Creek Mine, North Carolina. *Smithsonian Contributions to Paleobiology* 90: 233–365.

Paruk, J. D., D. C. Evers, J. W. McIntyre, J. F. Barr, J. N. Mager, and W. H. Piper. 2021. Common Loon (*Gavia immer*), version 2.0. In *The Birds of North America,* edited by P. G. Rodewald. Ithaca, N.Y.: Cornell Lab of Ornithology. https://doi.org/10.2173/bna.

Rizzolo, D. J., C. E. Gray, J. A. Schmutz, J. F. Barr, C. Eberl, and J. W. McIntyre. 2020. Red-throated Loon (*Gavia stellata*), version 2.0. In *Birds of the World,* edited by P. G. Rodewald and B. K. Keeney. Ithaca, N.Y.: Cornell Lab of Ornithology. https://ezproxy.sjcme.edu:2102/10.2173/bow.retloo.02.

Russell, R. W. 2018. Arctic Loon (*Gavia arctica*), version 1.2. In *The Birds of North America,* edited by P. G. Rodewald. Ithaca, N.Y.: Cornell Lab of Ornithology. https://birdsna.org/Species-Account/bna/species/arcloo.

Russell, R. W. 2018. Pacific Loon (*Gavia pacifica*), version 1.2. In *The Birds of North America,* edited by P. G. Rodewald. Ithaca, N.Y.: Cornell Lab of Ornithology. https://birdsna.org/Species-Account/bna/species/pacloo.

Solovyeva, D., J. D. Paruk, J. Tash, S. Vartanayn, G. Danilov, and D. C. Evers. 2017. Post-breeding densities and lake size partitioning of loon species in western Chukotka, Russia. *Contemporary Problems in Ecology* 10 (6): 621–31.

Sprengelmeyer, Q. D. 2014. A phylogenetic reevaluation of the genus Gavia (Aves: Gaviiformes) using next-generation sequencing. Master's thesis, Northern Michigan University, Marquette.

Uher-Koch, B., N. R. North, and J. Schmutz. 2019. Yellow-billed Loon (*Gavia adamsii*), version 2.0. In *The Birds of North America,* edited by P. G. Rodewald. Ithaca, N.Y.: Cornell Lab of Ornithology. https://doi.org/10.2173/bna.121.

Williams, T. D. 1995. *The Penguins.* Oxford: Oxford University Press.

2

Selected by Nature

A Tale of Two Birds

I lowered my arm off the front end of the boat, opened my hand underwater, and with a measuring stick confirmed what I had suspected: it completely disappeared 18 inches below the surface. How could any visual predator find prey under these conditions? In my more than thirty years as a naturalist, I have had many amazing encounters with wildlife and observed a lot of fascinating behavior, but this story ranks up there as one of my most memorable. In January 2012, I was doing winter loon surveys off the Louisiana coast in the Gulf of Mexico. My mind was struggling to comprehend how loons could find food in this environment. They are visual predators, and the murky water must have made it extraordinarily challenging to locate prey. We were in a series of channels, dotted with islands, when we spotted a bottlenose dolphin. Any dolphin sighting makes for a great day, but this one was acting peculiar—it was swimming toward shore and would then swim back to the middle of the channel and repeat the same maneuver. What was it doing? we were all saying to ourselves. We could see ripples of water near shore from fish racing away from the dolphin. It must be pushing the schooling fish to shallower water to increase its foraging odds. That's it, we shouted to each other. Amazing. Then we saw a Common Loon following in the wake of the dolphin as it moved toward shore. At first, we were skeptical, but the longer we observed the more certain we were that the loon was consciously swimming back and forth each time in the wake to get any fish the dolphin missed. The behavior did not look cooperative by any means; it simply looked like the loon was aware of what the dolphin was trying to accomplish and knew that its chance of catching fish increased when swimming in the wake. Now, that was really amazing.

We continued our survey to more open water, listening to the ha-ha-ha-ha-ha of the Laughing Gull and bird-watching along the way. It was a beautiful day; the late-morning sun allowed us to remove some outer clothing, and I was thankful for this opportunity to enjoy sunshine and warmth, knowing if I were back home, I would be in a down jacket. We were feeling fairly good about the morning, because in addition to observing the dolphin-loon interaction, we observed some Roseate Spoonbills. Living in Maine and being a rare visitor to the Southeast, I marveled and appreciated the stunning pink coloration of the spoonbills. It could not get much better, I mused under my breath. Then, in the distance we saw an enormous flock of birds, unlike anything I had witnessed before. There must be several thousand, someone shouted. We counted by groups of fifty, then by groups of one hundred, and estimated roughly three thousand individuals. They were mostly in a row, which stretched out for several football fields. The birds were quite large, all black, and their long bills, hooked at the tip, gave their identity away. I had observed hundreds of Double-crested Cormorants before, but never thousands—truly a sight to remember. I never imagined I could learn so much about loons by studying cormorants. As Mickey Mantle once said, "It is unbelievable how much you don't know about the game you have been playing all your life."

To Be a Loon or a Cormorant

•

A breeding Common Loon has stunning plumage. The upperparts are checkered with rows of white rectangles across a glossy black body. The dark, thick neck has white vertical bars, or lines, on its sides that form a distinctive necklace. The chest and belly are uniformly white and contrast with the darker upperparts. The eye is vivid red and stands out against the jet-black face. Pictures of loons grace many houses, calendars—even airport walls. These birds are almost universally adored. The Double-crested Cormorant is an equally exceptional diving bird, but pictures of them do not grace houses, calendars, or airport walls. The plumage of the cormorant is uniformly dark on top and bottom; there are no contrasts, no white checkered rows or white belly. They have no markings on the neck. Their eyes are yellow. The Latin name for cormorant is *Phalacrocorax auritus,* which translates into the eared, bald crow, or raven, presumably a disparaging term that summarized the opinions of the bird of early naturalists (such feelings and opinions about cormorants also exist in the present day). Both loons and cormorants are diving birds that live in freshwater and specialize in eating fish, but why is one adorned with a spectacular breeding plumage while the other is not? Since the days of Aristotle, we have come a long way in explaining patterns in the natural world; we know more. Rudyard Kipling's *Just So Stories,* which explain why zebras have stripes and leopards have spots, are entertaining but fall short of a scientific explanation. For that, we must look to Darwin. Natural selection, over eons, has shaped basic animal shapes, forms, and behavior. As a scientist, for years I was obsessed with the striking contrast between these two similar birds and wondered why the difference. Why should two highly successful birds with similar lifestyles have such dissimilar markings? I think I have an answer.

Loons and cormorants are dark colored, and so are other waterbirds, like penguins, puffins, and guillemots. However, a few noticeable birds associated with water, such as swans, gulls, and terns, are predominately white. Coloration in feathers is mainly due to pigments, and melanin is one of the main pigments in birds. Humans also have melanin; it is responsible for differences in our skin and hair color. The more melanin deposited into the feather follicle, the darker the feather; it is that simple. So, if a bird lacks melanin, its feathers will be white; if it has varying or moderate levels, its feathers will be light to dark gray; and if it receives lots of melanin, its feathers will be black. Melanin is a complex polymer that occurs in two different forms, eumelanin and phaeomelanin, and it is the former that is responsible for whites, grays, and blacks (the latter is responsible for browns, reddish browns, and tans). Have you ever noticed that the great majority of white birds have black wing tips? The American White Pelican, Snow Goose, Northern Gannet, and the great majority of gulls, terns, and albatrosses all have black tips. Now, have you pondered that essentially all diving birds are black? Cormorants, loons, penguins, razorbills, murres, and puffins are universally black on top. What is the connection between the two? Melanin does more than color the feather; it makes the feather stronger and more resistant to wear. The added strength likely comes from both the melanin granule and the additional keratin (the protein responsible for the strength of our fingernails and toenails) that gets deposited in feathers with melanin. Simply put, white feathers are structurally weaker than dark feathers (aluminum versus steel). Therefore, areas of a bird, like wing tips and tail tips, that are more likely to wear during flight, or experience abrasion from twigs or branches, are black. However, there is more to the story.

The first time I held a loon on my lap I could not help but notice the rigidity of the feathers; they were remarkably stiff. I had handled several bird feathers in my ornithology classes, and the wing feathers of a loon reminded me of the tail feathers of a woodpecker. A woodpecker uses its stiff tail feathers to support itself as it walks up the sides of trees; collectively the feathers act like a third leg. Why would the feather of a loon wing be so stiff? Birds have air sacs distributed throughout their body, and if they are fully expanded, residual air would make the dive appreciably harder. Imagine how more difficult it would be to dive if you carried partially inflated balloons tucked under your shirt. Stiffer feathers compress the body, squeezing those air sacs empty and thus making diving below water easier.

Birds make melanin from amino acids, which they obtain from the protein in their diet (insects, fish, birds, and mammals). It takes energy to synthesize melanin, and birds get that extra energy by catching and eating prey. But what if they didn't need to make it? That energy could be redirected to manufacture equally important biomolecules (hemoglobin to make red blood cells, for example). Why go through the trouble to produce melanin if you do not have to? Yet, nearly all diving birds around the world are black, not white, because the additional structural integrity of the feathers increases their underwater diving efficiency. There is at least one species of waterfowl, the Smew, that appears to be the exception to the rule. They dive to get insects and fish, and yet the males are predominately white in their breeding plumage. How can they get away with being white? Unlike the melanistic diving birds, which routinely descend to 100 feet or more, the Smew rarely descends more than 13 feet and instead makes numerous shallow dives (3 to 10 feet). Natural selection has converged on melanistic pigments in all these diverse and unrelated diving birds (representing five separate orders) because it aids in their diving ability and, ultimately, their survival.

Checkered or Plain Back

•

From an adaptationist (or selectionist) point of view, two things are clear: first, the prominent plumage markings on the back, or dorsal side, of a loon enhance either its survival or reproduction, and, second, the lack of markings on a cormorant in no way affects its ability to survive and reproduce. Why has selection favored striking markings in loons and not in cormorants? What is so different between the two similar species to warrant such a marked contrast? The answer lies in the fact that the nesting and breeding ecology of the two birds is very different. Loons are territorial: they defend a lake or a part of a lake, such as a cove, from other loons, and sometimes they compete for the opportunity to breed. These contests are sometimes intense, and occasionally a combatant will die. Cormorants, on the other hand, may squabble with neighbors but not with the same intensity exhibited by loons.

Generally, markings on an animal either enhance its ability to conceal itself, or they do the opposite: they advertise it. Stripes on snakes, zebras, and tigers are examples of the former, while the stripes on skunks and yellow jackets are examples of the latter. Spots serve the same purpose. Stripes and spots work in concert with the overall color of the animal, and its primary background is to serve either as camouflage or advertisement. First, let's consider the evidence for or against the camouflage idea. Loons nest on islands, or on the tips of peninsulas, to reduce contact with mammalian egg predators, such as raccoons and skunks. Both sexes incubate equally, but while incubating, loons are vulnerable to their main aerial predator, the Bald Eagle. I have observed a Bald Eagle attack and swoop down on a nesting Common Loon on more than one occasion. Clearly, nesting loons are vulnerable to such aerial attacks, and we can make a strong case that an incubating loon would want cryptic instead of advertisement coloration. We would be equally correct in assuming that any pattern that concealed incubating loons from eagles would be selected for over time, so that generations of loons would benefit from such coloration. However, the distinctive checkerboard pattern evident on breeding loons does not appear to camouflage them. Case in point, over the years I have flown aerial surveys to count ducks or eagles, and I can emphatically state that I can see the silhouette of a loon clearly from several hundred feet up in the air. From that distance it appears the white markings on the back of a loon help accentuate it, not camouflage it! Eagles have great visual acuity, and we can reasonably conclude that one perched high in a tree sees an incubating loon just as easily as I do from the air and that the checkered back on a loon does not camouflage it. So, we are left with the advertisement hypothesis to account for the checkered backs of loons. If this hypothesis is correct, then to whom are they advertising?

Quite possibly they are advertising to other loons. I offer that it is a form of intraspecific communication, in a way a type of advertisement. Consider that during practice, ballerinas wear black because dark colors bring out contrasts, enabling the instructor to detect misalignments of the arms or legs by the pupil more easily. During a performance, they wear white because it emphasizes continuous movement and thus the audience will not easily detect misalignments. The point is that black-colored birds can be seen over greater distances (as silhouettes) than white-colored birds (e.g., eagle versus egret). Birds that advertise themselves over long distances are predominately black, such as crows, ravens, and eagles. Black enables the greatest contrast with most backgrounds, and it shows the bird's shape exceedingly well. Where does the white come in? White contrasts maximally against black, reflecting all possible light, and thereby accentuating it. North American birds with pronounced black-and-white markings, such as the Bobolink, Lark Bunting, and Black-billed Magpie, typically reside in open habitats, not densely forested areas, to advertise their presence to others of their species. Loons also reside in open areas, lakes. I suspect loons are black and white to signal over long distances to other loons (intraspecific communication). What are they communicating?

A loon flying over a lake can simply look down and see if it is occupied by loons. If a lake is occupied, the flying loon does not have to waste energy by landing and investigating; it can fly to the next lake. By visually signaling to flying loons that the lake is occupied, a loon saves both the resident and the intruder time and energy. It is a win-win situation. Any mutation in loons that caused a change in its plumage pattern to increase long-distance communication would be selected for because it would increase either their ability to reproduce (spot a territory, a potential mate) or to survive (avoid a fight with a stronger rival). Loons are colored like skunks because they have a simple message to communicate to other loons: *I am here.*

What about the noncheckered backs of cormorants? The Double-crested Cormorant is the only freshwater breeding cormorant in North

America. All cormorant species nest in colonies on islands. Colonies tend to be large and easily visible. Thus long-distance communication plays less of a role for cormorants. They can easily find each other just by going to the colony. They have no need to search for a territory or a mate—everyone flies to the same island. When cormorants are breeding and sexually mature, the sides of their cheek and the base of their bill turn orange and are used in breeding displays, but that's about it. If we compare cormorants to other seabirds that nest in large colonies, such as puffins, razorbacks, murres, and guillemots, we find that all these species are uniformly colored on their dorsal side. This further supports the notion that the checkered pattern of breeding loons is an important part of their ability to communicate over long distances. Ultimately the ecology of the species (where they live, how they make their living, how their food is distributed in the landscape) usually dictates plumage patterns in birds. It turns out that cormorants appear dull for the very same reason that loons are brilliantly patterned.

Black or White Belly

The white belly, or ventral side, of a Common Loon is very noticeable because it contrasts maximally with the bird's dark upperparts, especially when a loon decides to lurch up and spread its wings. This is why loons are often seen from shore by homeowners. However, this marked contrast between the top and bottom of a loon is found on nearly every other diving bird. The pattern is so common in the animal kingdom that there is a name for it, countershading, and it serves an important purpose. In the late 1980s, biologists wanted to study the Marbled Murrelet, a small diving bird of the auk family of western North America, because their numbers appeared to be declining. Researchers began to catch and band individuals, but the birds were hard to catch because field-workers could not locate their nests. Thus, they had to catch them when they were feeding and resting on the open ocean. At first, they used a net gun, but that proved challenging. One group was so frustrated that they tried having U.S. Navy Seals in scuba gear swim under the birds and catch them with a net attached to a pole. This method was also unsuccessful. The Seals could not see the birds from below—it was as if they were invisible. This is the value of countershading; it serves as a type of camouflage. Against the light sky, the murrelets' white bellies are invisible, which provides protection from potential underwater predators, such as large fish or even sharks. As unbelievable as it may seem, I assisted in banding a loon chick in the Upper Peninsula of Michigan in 1992, and several years later, when it washed up on a beach in Virginia, a shark tooth was lodged in its body. Loons live in the open ocean for three to four months of the year (November to March), and having a white belly in this environment likely increases their chances of survival, just as it does for the other marine diving birds.

But if countershading aids in the survival of diving birds, then why don't cormorants have white bellies? Besides cormorants, there is another group of freshwater diving birds that are not countershaded, called darters. One species of darter, the Anhinga, occurs in the southeastern United States. It is also called snakebird because it swims like a serpent with its body completely submerged and only its long neck and head above the water. Cormorants and darters share a similar (and most unusual) behavior called "sunbathing," or "wing drying," in which they spread their wings and face their backs to the sun. By all accounts, they certainly look like they are sunbathing. Are they cold? Are they drying their feathers? Moreover, does this information help us explain why they are not countershaded?

Cormorants and darters are related, and similar groups of animals often share the same characteristics and behaviors. In this case, unlike other marine diving birds, cormorants and darters increase their efficiency swimming underwater by allowing their feathers to become wet, which decreases their buoyancy. In other words, their feathers are not water-repellant. It was thought at one time that when oil from a bird's preen (uropygial) gland was rubbed into feathers, it made them water-repellant. Instead, new data suggest that the oil helps keep the feather tips rigid, and the tight overlapping of feathers at the microscopic level keeps the plumage waterproof. Cormorants and darters lack an oil gland altogether, so they need to dry their feathers on land, and they spread their wings to do so. Because darters have unusually low metabolic rates and unusually high rates of heat loss from their bodies, their challenge is to stay thermoregulated, especially when it is cool, cloudy, and windy. They assume the spread-wing posture to maximize solar absorption, and their dark bellies retain more heat relative to a white belly. The dark belly aids in solar absorption and accelerates the drying of the feathers, which means the bird can get back in the water to resume foraging.

An adult Double-crested Cormorant in a "sunbathing" pose.
Photograph by Colin Durfee. Licensed under CC-BY-2.0.

Since both of these birds also fly and perch up on a tree limb to spread their wings, they inevitably regularly brush against limbs, branches, and twigs. Because the pigment melanin makes feathers stronger and more resistant to wear, having a dark underside offers increased protection to their belly feathers. Cormorants spend less time in the water than loons, and thus the lack of countershading may not be as important for their survival. When a cormorant *is* in the water, it is busy foraging, usually for thirty minutes at a time, and when it is done, it flies to land to preen and rest. When not incubating, a loon spends nearly all its time in the water, so countershading may be more important for its survival than it would be for a cormorant.

Necklace or No Necklace

•

Why do some birds have markings on the neck while others do not? Although I am uncomfortable applying human culture to patterns observed in animals, this is one of those rare times it might be instructive to do so. Think about it: why do people wear objects on their neck? In most situations, it is to draw attention to the artifact or jewelry they are wearing. Other animals, besides humans, likely have markings on the neck for the same reason. A uniformly colored neck with no markings does not stick out from the rest of the body. It makes it much more difficult to tell if the neck is held high or low, for example. But markings on the neck do the opposite. They serve as a form of communication and allow the receiver to better interpret the intentions of the sender, and vice versa. The position of the neck tells a great deal about a bird. Is it alarmed, nervous, or ready to take flight? A dominant individual holds the neck up high; a subdominant individual holds it low. Neck markings also allow communication to take place from a distance. Assessment of how an individual is feeling can be determined from a distance, and the proper behavioral response can be mediated. Confrontations can be avoided.

Take the neck markings on geese, for example. Canada Geese and Brant, two black-necked species of geese that breed in North America, have distinct markings on their necks. Snow Geese and Ross's Geese, two all-white species of North American geese, lack markings on the neck. Once again, we note that the breeding ecology between black-necked and white-necked geese is different. Canada Geese and Brant are solitary nesters that defend an area around their nest from other rivals. Conflicts occur. How many times do you recall observing a Canada Goose lowering its head and charging another goose (or yourself!)? On the other hand, Snow Geese and Ross's Geese both nest in large colonies. They defend only a small area immediately around the nest. Conflicts occur, but they are settled quickly and with less commotion. Competition for nesting spots is less intense.

Loons are strongly territorial and often engage in skirmishes with other loons in which they often have to interpret the intention of the other animal (much like two dogs meeting for the first time). The necklace allows the intention of the individual to be better interpreted by other loons. If the bird is submissive, its head will be low in the water, and the necklace will rarely be exposed. If the individual is feeling dominant, its head will be held high, the necklace fully exposed. This might convey the state of the intruder to the resident. Should the resident prepare for a battle? We can see how under these conditions a necklace aids in intraspecies communication, which brings us back to the lack of neck markings on the cormorant. Cormorants are colonial nesters;

individuals are clustered together and are not assessing each other from great distances. Other colonial nesting marine species, such as puffins, murres, and frigatebirds, also lack neck markings. There have been few empirical studies on the topic, but I suspect that neck markings on colonial nesting waterbirds are less critical to interpreting the intention of their neighbors than they are for loons. To the adaptationist, this suggests that neck markings are important in species that rely on or use long-distance communication frequently.

Common Loons lose their necklace during the winter, which is one of the reasons they are hard to identify at that time of year. The behavior of the loon also changes dramatically between summer and winter. Loons wintering in the Gulf of Mexico or in Monterey Bay are generally solitary and do not defend a territory from other loons. There are no confrontations, no battles for territorial rights, no squabbles—they just go about their business of trying to find food. The importance of their necklace is diminished since they are less worried about understanding the intentions of their neighbors.

Red or Yellow Eye

•

Many books about loons claim that their red eyes are an adaptation. Here again, a comparative approach is useful in interpreting the function of the red eye. Most birds have a dark iris, mostly brown, and some have a colored iris, usually white, yellow, orange, or red. For example, the Red-eyed Vireo and White-eyed Vireo are named aptly. Some biologists think that the color of the eye allows members of the same species to recognize each other (a species-specific signal). But remember, loons recognize each other at a long distance, so they likely do not need to see a red eye to tell them they are looking at another loon.

In some species, iris color differs between the sexes. For example, male Brewer's Blackbirds, Surf Scoters, and Northern Shovelers have yellow eyes, and the females of each species have brown eyes. In loons, there is no difference in eye color between the sexes. This is not surprising, since the great majority of birds show no difference in eye color between the sexes. Eye color can and often does vary between juvenile and adult birds. In many hawks, juveniles have brown eyes during their first winter and by their second year typically take on the iris color of the adult (yellow or red). This developmental change in eye color helps birds who have no obvious plumage differences, because it aids in distinguishing between juveniles and adults in the population. Loons are like many hawks in this respect in that juveniles have brown eyes in their first year and acquire the red eye in their second year. During winter, although the adult loon's eye is red, it is much duller in color.

Other dark-headed waterbirds, such as the Black- and Yellow-crowned Night Herons, and both the Horned and Eared Grebes, also have red eyes. This suggests there might be selection for red eyes in animals with dark heads. Physicists instruct us that greens, blues, and oranges do not contrast against black as well as red. Sociologists tell us that red and black invoke an emotional response, at least in humans, and presumably in other animals. The color combination of red and black gets our attention. Fashion designers are aware of this effect, and that is why red and black dresses are popular. Biologists have also observed the red and black combination on many male songbirds (e.g., Scarlet Tanager, Magnificent Frigatebird, and Northern Cardinal). Therefore, it seems selection might favor birds of breeding age with red eyes because the red eye evokes a strong signal in the opposite sex. But not all dark-headed waterbirds have red eyes. For example, our beloved Double-crested Cormorant has yellow eyes. As do scoters, Hooded Mergansers, and Canada Geese. Phenotypic similarities and differences exist between species, and scientists look for patterns to gain insight into what selective pressures might be driving evolutionary success. In the case of the Common Loon and Double-crested Cormorant, both red and yellow eyes are found in mature birds with dark heads. Selection appears not to favor one over the other—an important reminder that animal diversity is the product of innumerable constraints and possibilities.

The red eye may serve as a sexual maturity signal, but could it also aid in underwater foraging success? The logic goes like this: at about 15 feet underwater, red appears gray or neutral, so loons with red eyes would be more successful foraging for food than loons without red eyes because they would be less likely to be detected. What are the data for this? Well, there are none; it is all theoretical. If loons having red eyes enjoyed better foraging success (and consequently increased their chances of survival), we would expect other, similar fish-eating birds would benefit from having red eyes too. If red eyes increase foraging success relative to individuals with different color eyes (say, yellow), selection would nudge and pull eye color in that direction, unless there were genetic constraints. The comparative data suggest to me that the red eye is probably not crucial to a diving bird's foraging success. Moreover, when we compare red-eyed and non-red-eyed piscivores, an interesting pattern emerges. The noncolonial nesting species (loons, mergansers, and Western Grebe) have red eyes, and the colonial nesting species (cormorants, murres, and razorbills) lack them. We know that loons often rely on long-distance communication, and colonial nesting species presumably use it less. This suggests that the red eye is involved in communication and may be a sign of breeding readiness, potentially selected for by mates. During the winter, when loons are not breeding or spending time with their mates, their eyes are a dull red.

Lastly, pigments are mostly responsible for eye color in birds. Melanins produce dark eye colors, and carotenoids produce yellow eyes in many diving ducks. Another pigment, pterin, is synthesized by birds and responsible for the striking red eye colors (hawks, doves). Pterins can also produce striking yellow eyes (Brewer's Blackbirds). Less than a third of the breeding bird species in North America have had the sources of their eye color discerned (loons fall into that category). Before we can infer natural selection as responsible for shaping eye color in birds, we need to know whether these pigments are ingested or synthesized, from where they are obtained, and the genetics behind their production. Until more empirical data are obtained, explanations for the loon's red eye will remain a mystery.

Why Loons and Not Cormorants?

•

Literature in cognitive psychology discusses how quick we are to judge an animal, person, even a movie, and classify it. As humans, we crave order; we search for a pattern (even when there is not one) and feel content and relaxed when we have established one. The human brain is very good at categorizing information. When we can quickly put something in a mental box, we feel good, and when we cannot, it causes us consternation. Consider this in terms of wildlife. Animals that are black and white are easy to recognize, and we know where they go in the box. We like that. We like loons because everyone can identify them.

Cormorants lack distinctive markings (necklace, red eye) and are less easy to identify. They do not fit neatly into a box and therefore suffer the fate of human bias. We prefer birds that are easy to pick out, with bold patterns or bright colors. But is it as simple as that? Partly, yes, and partly, no. We notice and respond to loons because they vocalize loudly and in a rhythmic pattern that evokes an emotional response in us. It hits a cord, so to speak. Cormorant vocalizations do no such thing. Lastly, I think we have more empathy for loons than for cormorants because we often see the former as a family unit, while the latter we do not. Because of all these distinctions, cormorants run a distant second to loons in the two-bird popularity contest.

Further Reading

Caro, T. 2009. Contrasting coloration in terrestrial mammals. *Philosophical Transactions of the Royal Society of Biological Sciences* 364 (1516): 537–48.

Dorr, B. S., J. J. Hatch, and D. V. Weseloh. 2014. Double-crested Cormorant (*Phalacrocorax auritus*), version 2.0. In *The Birds of North America,* edited by A. F. Poole. Ithaca, N.Y.: Cornell Lab of Ornithology. https://doi.org/10.2173/bna.441.

Frederick, P. C., and D. Siegel-Causey. 2000. Anhinga (*Anhinga anhinga*), version 2.0. In *The Birds of North America,* edited by A. F. Poole and F. B. Gill. Ithaca, N.Y.: Cornell Lab of Ornithology. https://doi.org/10.2173/bna.522.

Hill, G. E. 2010. *Bird Coloration.* Washington, D.C.: National Geographic Books.

Klein, T. 1989. *Loon Magic.* Minocqua, Wis.: NorthWord Press.

Lovette, I. J., and J. W. Fitzpatrick. 2016. *Handbook of Bird Biology.* 3rd ed. West Sussex, UK: John Wiley and Sons.

Paruk, J. D., D. C. Evers, J. W. McIntyre, J. F. Barr, J. N. Mager, and W. H. Piper. 2021. Common Loon (*Gavia immer*), version 2.0. In *The Birds of North America,* edited by P. G. Rodewald. Ithaca, N.Y.: Cornell Lab of Ornithology. https://doi.org/10.2173/bna.

Peterson, R. T. 2008. *A Field Guide to Birds of North America.* New York: Houghton Mifflin, Harcourt Publishing.

Rudd, M. M. 2005. *Rare Bird: Pursuing the Mystery of the Marbled Murrelet.* Seattle: Mountaineers Books.

Terres, J. K. 1980. *Encyclopedia of North American Birds.* New York: Alfred A. Knopf.

Wires, L. R. 2014. *The Double-Crested Cormorant (Plight of a Feathered Pariah).* New Haven, Conn.: Yale University Press.

Zahavi, A., and A. Zahavi. 1997. *The Handicap Principle: A Missing Piece of Darwin's Puzzle.* Oxford: Oxford University Press.

3

What a Drag!

THE INNER WORKINGS OF A MASTER DIVER

In 1989, I committed a week of my summer to helping an undergraduate college friend, Dave Evers, catch and band loons as part of his master's thesis at Western Michigan University. I met Dave at Seney National Wildlife Refuge in Michigan's Upper Peninsula. Catching a loon requires going out at night in a small boat, armed with a bright light and a long-handled net. As we approached a family of two adults with two chicks that night, one of the parents dove below the surface. From the boat we tracked it with the light as it swam underwater. I will never forget the powerful, synchronous movements of its legs, and the amount of distance it traveled between strokes. I was astonished to witness this aspect of a loon, a view few people experience. I was struck by its shape, narrow and sleek under the water, not as large and bulky as it appears on the surface. Then, with a quick stroke and twist of its body, the loon dove toward the bottom of the lake, away from the light.

Loons are amazing divers, and their anatomy has been modified over time to reduce drag and minimize resistance. The average person does not think much of drag unless they are biking into a heavy wind or towing a trailer cross-country. Anything, be it a car, a bird, a fish moving through a fluid, be it air or water, experiences resistance to its forward movement because of its interaction with the surrounding fluid. This resistance is known as drag (called aerodynamic in air, hydrodynamic in water). Drag increases with the density of the fluid, and since water is more than eight hundred times denser than air, a lot more drag is created swimming than running or flying. The goal of any swimming animal is to move through the water efficiently so its energy is conserved. For large swimmers, like loons, a teardrop is the best hydrodynamic shape when moving through the water because it produces the least possible resistance. Think of a canoe or a kayak: the front is narrow, then widens in the middle before narrowing again at the end. If the front or end were wider than the middle, additional drag would develop. Shape matters a lot, so much so that three different evolutionary lineages—fish, mammals, and birds (tunas, dolphins, and loons, respectively)—have all converged on the teardrop shape. Drag is also affected by the surface of the object; a smooth surface will create less drag than a rough one (e.g., polished glass versus sandpaper). This is why competitive swimmers shave their bodies or, in the case of the 2004 and 2008 Olympics, wear full bodysuits.

Skeletal Modifications

Loons are master underwater swimmers, but they did not get this way overnight. Natural selection has modified their skeletal system in various ways to reduce drag and make them more efficient in the water. Some of the ways will be obvious, but a few may surprise you. First, *density.* If you are attempting to get to the bottom of a lake to forage for food, it helps to get there faster so you can spend more time searching and pursuing prey. Added weight is a good thing if you are a deep-diving bird—just ask scuba divers why they add weights to their wet suits. The skull of a loon is unusually thick and heavy for a bird; this allows it to descend with less effort. Also, many of a loon's bones are solid rather than mostly hollow, or pneumatic, as found in most birds. The solid dense bones add weight, which makes it easier for loons to descend, thereby saving energy. These dense bones increase a loon's density more than one would expect for a bird its size. For example, a Bald Eagle is 20 to 25 percent larger than a loon yet weighs about the same (8 to 11 pounds).

Second, *shape.* This is the area natural selection has modified the most in loons. It is observable in literally every part of their anatomy, from their skull, to the shoulders and hips, to their legs and feet. To reduce drag, a loon's skull is streamlined, with a low and long profile, widening only at the base. This shape allows water to flow up and over the skull with minimal drag. Among diving birds, loons have one of the narrowest shoulder (pectoral) girdles. Their sternum, or breastbone, is long and narrow, and their keel is shallow. The shallow keel keeps the profile of the loon thin and sleek, minimizing drag. A shallow keel, however, does not provide much room for breast muscle attachment, which means they have less power to generate the lift necessary for takeoff. This is an example of an ecological trade-off. It is hard to be really good at both swimming well underwater and taking off into the air. Since loons spend nearly all their time in the water, and their first impulse to avoid danger is to dive, not fly (as it is in most birds), it seems that selection to increase the loon's underwater efficiency would be more advantageous than increasing takeoff efficiency.

Birds that spend more time in the air have enlarged keels; in fact, hummingbirds and swifts, two groups of birds noted for their mastery of flight, have greatly enlarged keels for their size. A keel is a ridge on the breastbone (or sternum) that serves as a major site for flying muscle attachment. Moreover, flightless birds, such as ostriches, emus, and kiwis, lack a keel altogether. Because such patterns are repeatedly observed in the animal world, the concept of trade-offs is well established. The hips of a loon are astonishingly narrow, and the lower leg is flattened, both designed to reduce drag. A loon pelvis is 33 percent narrower than that of a grebe, for example, considered by many to be one of the best-adapted diving birds in the world. The loon's lower leg is shaped like a canoe paddle, maximizing thrust on the downstroke and minimizing drag on the recovery stroke. For the downstroke, the flat and broadest part of the lower leg is twisted, and the toes are spread wide apart, enabling the loon to push the maximal amount of water. But on the recovery stroke, a simple twist dictates that the narrowest portion of the bone is exposed with the toes tucked behind the foot bone, minimizing the resistance.

Third, *length.* For a bird its size, two parts, the arm and femur, are noticeably shorter, and three parts, the ribs, calf, and toes, are noticeably longer. Based on the relationship between body length and wingspan, a bird 32 inches in length should have wings roughly 60 to 68 inches long. However, the Common Loon's wingspan is 50 to 58 inches long, or 20 percent shorter than predicted for its body size, and the wingspan of a similarly sized Short-tailed Albatross is 84 to 90 inches, or 40 percent longer than predicted. It is logical that selection would favor longer wings in a bird that makes its living gliding, like an albatross, but why would a loon's wings be selected to be shorter? When a loon swims underwater, the wing is folded and pressed tight against

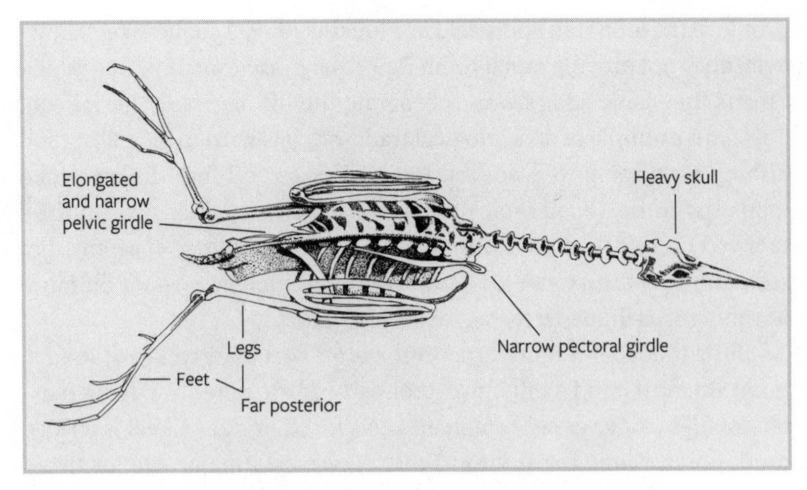

Full articulated skeleton of a Common Loon.
Modified from McIntyre, *The Common Loon* (1988).

the body (remember, they are foot-propelled divers, unlike penguins, which flap their arms to generate thrust). The body of a loon is 30 to 32 inches long, and if the folded wing were longer than the body, it would protrude, causing eddies and unwanted drag. It is possible that selection has shortened the loon wing (25 to 29 inches) while optimizing its length to increase the bird's efficiency for diving but, alternatively, has made it harder to attain and maintain lift in the air. Furthermore, the femur (thighbone) of a loon is remarkably short—the shortest of any diving bird in North America—the upper leg being tucked up next to the rib cage within the abdominal skin, effectively eliminating drag.

The ribs of many birds are enlarged, probably a consequence of needing to support muscles that are associated with flight. Loons have ten ribs, nine of which articulate with the spine, and eight with the sternum. The tenth rib is free floating and not attached to the sternum or the spine. Most birds have an unusual long bony projection from one rib that overlaps with the neighboring rib, called an uncinate process. The uncinate process is thought to strengthen the rib cage to withstand the tremendous compression forces generated during flight. Loons have an uncinate process, but theirs is even longer than normal since it reaches across two adjoining ribs. There are a few theories that might explain why they would need this extra strengthening, though more definitive studies are needed to verify them. It is possible, for example,

The streamlined shape of a loon underwater. Photograph copyright Daniel Poleschook and Ginger Gumm.

that loons put tremendous stress on the rib cage during flight because of their weight and rapid wingbeat (greater than 200 per minute). Another theory relates to the fact that loons are deep-diving birds. Loons diving 200 feet deep experience roughly 1,000 pounds of pressure per square inch, so the longer uncinate processes in loons may reinforce the rib cage to withstand such force.

The lower leg of the loon (tibiotarsus) has a huge spike coming off it, called the cnemial crest, essentially lengthening the bone. This additional length provides space for the calf muscle, which is used to power the foot stroke. A loon's toes, or digits (phalanges), are elongated and positioned wide apart. Lengthening and spacing the toes apart increases the surface area of the web. A large webbed foot increases the amount of water that can be pushed, effectively generating more thrust. The bones at the base of each digit have a small protrusion that causes all the digits to extend together. Loons also have one digit that is short and unwebbed and appears to serve no apparent function. Perhaps it is vestigial, like the human appendix, and serves as a reminder that the loon is a descendant of a bird that spent more time on the land.

Fourth, *angle of attachment.* The upper-lower leg attachments of the loon have been modified to permit a greater rotation of the joint (270 degrees versus 180 degrees for most birds). This increased arc allows loons to extend their feet far forward, much like a competitive

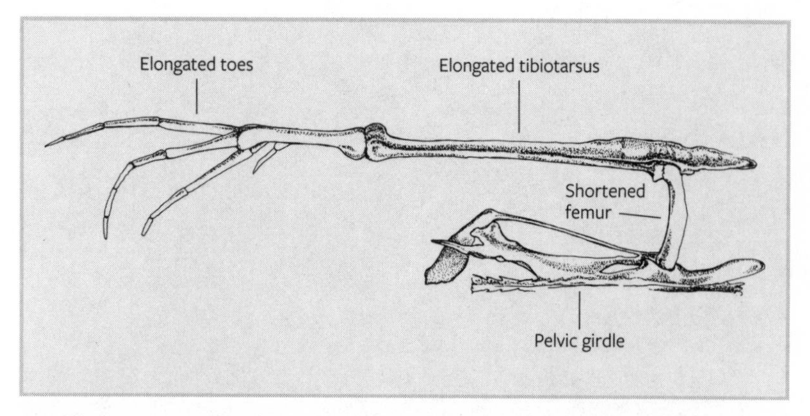

The lower extremity of a Common Loon. Note the short femur perpendicular to the pelvic girdle. Modified from McIntyre, *The Common Loon* (1988).

swimmer, effectively grabbing more water to be directed toward the rear, propelling them forward. A pattern emerges: diving birds that use foot propulsion are found in freshwater environments and have less developed keels and shoulder joints (grebes, mergansers); diving birds that use their wings for propulsion are found in saltwater environments and have more developed keels and shoulder joints (puffins, murres). A diving bird relying on wing propulsion in a lake with underwater plants would more likely have its swimming stroke impeded or its wings entangled than would a foot-propelled diver.

Muscles and Tendons

•

Just as the skeletal system of a loon shows modifications, its muscular system has also been shaped by natural selection. Muscles do work, they produce movement, and they span the joints of the body. A larger muscle can do more work with greater endurance than a smaller one. Muscles are expensive to maintain because they require lots of oxygen and glucose; thus, organisms must use muscles efficiently. Because loons spend considerably more time swimming than flying, we can reason that the muscles associated with each have been modified but in opposite ways. Relative to other birds, muscles used for diving and swimming would be enlarged, and those used for flying would be smaller. Loons exhibit this pattern. If we compare a loon with a wing-propelled diver, such as a puffin, we see striking differences in muscle size. The breast muscle of the puffin, for example, is much larger than a loon's (relative to adjacent muscles). Similarly, the muscle that rests on the back of the bird and is responsible for the upstroke (the supracoracoideus) is enlarged in puffins and underdeveloped in loons.

There are at least ten muscles, called fixators, that attach a loon's leg to its hip. The leg needs to be anchored well to the hip for obvious reasons, and probably even more so in a diving bird. As in nearly all birds, a loon's hip socket is shallow, and there is little space for the head of the femur to attach to the hip. Because of this odd bit of anatomy, the fixators in loons are highly developed. Fixators function not so much in movement or muscle contraction but mostly to keep the joint immobile. Muscles typically have tendons, or sheaths, at the point of origin and insertion because tendons are relatively inert, strong, and compact and are ultimately a lot less expensive to maintain than muscles. Muscle fibers have more volume and higher metabolic rates, require more blood, and are more expensive to maintain. Natural selection, being the parsimonious reaper of extravagance, ensures that tendons will replace muscle tissue whenever and wherever it is possible to do so. With that background, the die has been cast. Most of a loon's muscles that serve as flexors of the hip are fleshy throughout. This is uncommon in birds and suggests that the fixators are vitally important in loons and other foot-propelled divers, so much so that few of them have tendons at their origin or point of insertion. You might pull a drumstick from a turkey, but I doubt you would have the same success pulling a leg from a loon.

To be successful predators, loons need powerful jaw muscles that can reach out, grab, and hold on to fish. The musculature associated with the bill, or mandible, in loons is highly developed. It is interesting to compare the jaw muscles of loons to those of mergansers, another group of diving and piscivorous birds. Scientists calculating the force at which loons and mergansers can adduct and retract their jaws found that loons have 1.5 and 2.2 times more force than mergansers in their adductors and retractors, respectively. After having my finger stuck inside the beak of a loon, believe me, I can attest to the strength it possesses. Loons may have developed this additional force because they do not possess serrated, or toothed, cutting edges on their thin mandibles or an enlarged nail at the tip of the bill. Mergansers may not need added musculature because they have serrated beaks, which aid them in holding on to fish. Which would you rather have if you were catching fish, a beak with serrations or one without? My guess is you would choose a serrated beak. So, why is the loon's beak nonserrated? One idea is that serrated beaks evolved after loons went on their own separate evolutionary trajectory, and recent genetic analysis shows that might be the case. With that door closed, loons evolved in ways that would serve them well in catching and holding on to fish, hence larger jaw muscles. They also developed an elaborate tongue and associated musculature, both of which aid in manipulating and swallowing food. Few birds (except for hummingbirds and woodpeckers) have a more developed tongue than the loon.

Loons and a few select other species of birds (e.g., shorebirds) can move the upper bill and the lower bill independently, a motion called cranial kinesis. In most birds the upper bill is fused to the skull. Think of all the birds you observe at your feeders (jays, cardinals, goldfinches), and notice how only the lower bill is repeatedly moving up and down to crack seeds. There is a hinge in some birds at the junction of the base of the bill and the maxillary bone that allows the upper jaw (and bill) to open rapidly. How does this movement help loons catch fish? Imagine you were trying to catch a fish with a pair of chopsticks but could only move the lower one against the nonmovable upper one? The gap might be large enough for most fish, but if you wanted to improve your foraging success, you would want to be able to move both chopsticks, right? This ability to move both the lower and upper bill (not unique to loons) appears to be an adaptation that greatly aids loons in foraging for prey.

Modifications of muscles are, for the most part, a change in size or shape and in their extent of origin, which is typically widely spaced over the bony surface. These modifications may decrease or increase the power and effectiveness of the muscle. Rarely is there a change in the point of insertion. The point of insertion occupies a narrow line where the force of the muscle is concentrated. This cannot vary greatly, for if it does, the action of the muscle will be altered. Yet, this has occurred in loons. The femorotibialis muscle originates on the femur and inserts on the tibia and is responsible for the extension of the lower leg. Extension increases the angle of movement, and flexion decreases it. The femur of a loon is very short. A loon swims with its lower legs, not with its upper legs, so what has happened to the musculature of the femur? Has it atrophied? Shortened? No, it has done something radically different: it has changed its point of insertion. Extension of the femorotibialis is no longer possible in loons because instead of inserting on the head of the tibia in front of the joint, it is inserted on the medial and lateral side of the knee! The femur has lost its ability to extend the tibia but instead can rotate it side to side.

Finally, let's examine one other modification in the loon, the patagial tendon. The patagial is found in the arm, where it connects the shoulder/upper arm bone to the wrist. Its function is to keep the leading edge of the wing stiff since it receives lots of force from air moving over and under it, especially at high speeds. It assists in gliding, but in loons it is enlarged and potentially serves another purpose: it keeps the wing folded and appressed to the body. I suspect that the increased thickness of the patagial tendon assists in keeping the wings close to the body on dives, which greatly reduces drag and friction.

The Physiology of Divers

•

Physiology attempts to understand how an animal functions. Ecophysiology attempts to understand how living organisms adjust to their environment at the molecular, cellular, organ, and organ system level. For example, what do animals do if they are too warm, or too cold? Every animal has its own comfort, or thermoneutral zone, and above or below that temperature they begin to feel uncomfortable. They need to thermoregulate. Most dogs dig a hole or pant on a hot summer day. In birds, the hypothalamus regulates body temperature and initiates a series of events to either warm or cool the bird so it can get back to its thermoneutral zone. Besides thermoregulation, physiologists often investigate a process called osmoregulation, which explains how the body regulates its salt (ionic) balance. Osmoregulation is critical for loons since they spend several months of the year on the ocean. Diving birds like loons must contend with many such physiological challenges.

The environment—a desert or a humid tropical forest, for example—dictates thermal stress on an animal. A loon's primary environment is the water, which is a cooling medium because water conducts heat from an organism faster than air. Thus, the primary challenge facing a loon is staying warm. Although putting on a sweater is not in the loon tool kit, they do have an inner layer of feathers that acts almost as a layer of clothing. Because these feather "sweaters" are zipped tightly, they are also highly waterproof (think Gore-Tex). This structural adaptation goes a long way toward keeping loons warm. Ornithologists previously thought it was the oil from the preen gland that waterproofed the plumage of birds, but a number of studies have shown that it is actually the zipping and overlocking mechanism of bird feathers (barbs and barbules) that repels water. This is why birds, and loons in particular, are constantly preening and keeping their feathers free of dirt and debris that might interfere with the feathers. Because the ventral side of the loon is in constant contact with water, this area needs lots of attention, and that is why we often observe loons rolling over on one side to preen their bellies.

The overlocking feathers "zip" and buffer the loon from the cooling effects of water (which is often roughly 30°F lower than their body temperature). But what about the legs, which are unfeathered and also in constant contact with it? Many loon watchers have noticed that loons will sometimes ship one foot and place it on their backs, while keeping the other in the water to rudder. Most likely they are doing this to thermoregulate. Shipping one foot is a behavioral mechanism to reduce heat loss, but could a loon employ a physiological mechanism as well? Our feet get cold when our blood vessels constrict, reducing the flow of warm blood from our hearts. But this constriction leads to another problem when the cold venous blood from our feet (about 45°F) returns to our core (98°F)—it chills us. This chilling could be a serious problem for a bird that spends a lot of time in cold water, especially in the winter. Loons solve this problem with a heat exchanger.

A heat exchanger works when two surfaces are in proximity to each other and the warmer surface can transfer its heat by conduction to the cooler surface. Many marine birds (puffins, murres) have their descending arteries in proximity to their ascending veins so heat transfer can occur, warming the venous blood before it returns to the core. Humans' descending arteries and ascending veins are far enough apart to prevent significant heat transfer, and that is why we get chilled when our feet get cold.

Heat exchangers come in two types, rete (which means network) and non-rete (or non-network). In the rete system, the distal arteries and venules are more finely branched and have more connections than in the non-rete system. Both types transfer heat effectively, and though some lab studies have shown that the rete system exchanges heat more efficiently than the non-rete system, others have found no appreciable difference between the two. It does appear that the rete system may dissipate heat faster than the non-rete system. Loons and the great majority of perching birds and gulls have the non-rete system, while many diving birds around the world (e.g., grebes, cormorants, and all diving ducks) have the rete system. Because both systems have evolved many times independently, taxonomists do not use this trait to infer historical relationships among different groups of birds.

Heat Dissipation: Keeping Cool

A loon is rarely in danger of heating up, but when it gets out of the water to incubate its eggs, for example, the potential does exist. Remember, a dark-colored bird such as a loon will heat up faster than a swan. Birds, like dogs, lack sweat glands, and although they do not pant, they do something similar, called gular fluttering, vibration of the throat membranes and quick movements of the tongue to dissipate heat. Panting is more efficient because a dog has a thicker, larger, and more mucous-covered tongue than a loon. The gular region is the upper throat of a bird, a skin pouch, which is more developed in some birds (owls, cormorants) than others (robins, goldfinches). When fluttering, a loon opens its mouth and vibrates the thin membranes of the throat. This movement increases blood flow to the area, further dissipating heat loss through evaporation. Though gular fluttering is effective, at times it cannot keep up with the heat gain, and the loon has to take a dip in the water, much like you or I would on a hot summer day.

Given a loon's need to keep cool, a rise in temperature in the loon's environment due to climate change could affect its everyday behavior and impact breeding success. One behavior that might be affected is the amount of time they spend incubating eggs. In the 1970s, during research for her classic book *The Common Loon: Spirit of Northern Lakes,* Judy McIntyre noted that loons rarely left the nest unattended (99 percent of the time the nest was covered by a parent). Twenty years later I did a similar study and found that the average time loons remained on their nests in 1995 and 1996 was 94.5 percent. The summers of 1995 and 1996 were hot, especially June. In fact, the summer of 1996 in northern Wisconsin was the hottest on record up to that time (we have had warmer weather since). The incubating loons I was studying over those two summers were clearly heating up. They spent more time gular fluttering and would also leave the nest to seek refuge in the water. I frequently observed loons getting off the nest for six minutes at a time throughout the day. Leaving the nest unattended exposes loon eggs to predators, such as gulls, crows, ravens, and eagles. Yes, eagles. A couple of years ago I viewed a video feed captured by the Biodiversity Research Institute of a Bald Eagle cracking and eating an unattended loon egg!

While banding loons, biologists have noticed that sometimes the foot becomes very warm. I once placed a copper wire on a loon's leg, and it read 120°F. A loon can shunt warm core blood to its feet, the only part of the body not covered in feathers, and thereby quickly dissipate heat to its environment. Here is how it works. When you or I are overheating, our hypothalamus detects this rise in temperature, and to correct it, it sends out a hormone that opens our blood vessels. Dilated blood vessels are close to the surface, and this is a great way to lose heat fast to the environment. This is the reason our faces are flush on a hot summer day. By shunting lots of warm blood to their feet, loons can effectively beat the heat (though perhaps not as well as diving birds with the rete system).

Too Much Salt

You may be familiar with lines from Samuel Taylor Coleridge's famous poem "The Rime of the Ancient Mariner": "Water, water every where, / And all the boards did shrink; / Water, water every where, / Nor any drop to drink." The reason adrift sailors cannot drink ocean water is because it would dehydrate them more than if they drank nothing. Seawater contains a higher concentration of salt than what our kidneys can process; when seawater is ingested, the body tries to dilute the excess salt by pulling water from our cells, effectively shrinking and dehydrating them. Eventually, the cells die, organs fail, and, without medical assistance, death occurs. Loons spend the winter in marine environments and are frequently exposed to salt in their diet. A bird's kidney is able to excrete salts in a concentration only about one-half of that found in seawater. So unless it has a backup plan, the bird will become dehydrated. Yet, marine birds like puffins and albatrosses handle the additional salt load in their diet just fine. They manage this with a specialized gland located above each eye that produces a highly concentrated salt solution (1.5 times that of seawater). The glands are paired and connected to the nasal openings on the bill, where the salt solution weeps out. Just like marine birds, loons have this salt gland, which gets activated when needed (under hormonal control) and shrinks when not in use, such as during the breeding season. It takes energy to use and maintain the salt gland, energy that could be used for putting on fat, growing new feathers, and pursuing prey. But it is vital to the loon's survival, especially when wintering in marine habitats where exposure to oil spills (which can deactivate the gland) can be so devastating to marine birds.

How Long Can a Loon Stay Underwater?

I have been recording dive times for both breeding and wintering Common Loon populations for more than two decades. Dive times vary by water depth and clarity, type and abundance of prey, and prey behavior. Most breeding loons nest in bays or coves of lakes that are generally shallow, and forage primarily in the littoral zone closer to shore, where the water is relatively shallow (less than 15 to 20 feet). The majority of these dives are between thirty-two and thirty-six seconds. Kevin Kenow, a research wildlife biologist with the U.S. Geological Survey, has found that migrating loons stopping over at Lake Michigan dove into water over 180 feet deep and remained submerged on average for more than two minutes. At Lake Jocassee, a reservoir more than 300 feet deep nestled in the Appalachian Mountains of South Carolina, I have recorded dive times of more than three minutes (with some longer than four minutes). But how deep can a loon dive, and how long can it stay underwater?

Oxygen consumption places a fundamental constraint on how long any diving bird (or mammal) can remain submerged. The longer a bird stays underwater, the more challenges it experiences. For example, oxygen supplies become limited. Oxygen is important in the production of a molecule called adenosine triphosphate, or ATP, and without ATP, a cell will die. Unfortunately, cells cannot shuttle ATP between cells. Some cells can go without oxygen for a short period of time, but if they do, they begin producing lactic acid. The problem with lactic acid is that it hinders performance (i.e., muscle contraction, leading to muscle fatigue) and can lower blood pH. And there is an additional problem: the longer an animal stays underwater the more carbon dioxide accumulates in the blood because it is not being exhaled. Carbon dioxide circulating in the blood joins with water to produce carbonic acid, which can lower the blood pH even further. A shift in blood pH can lead to a suite of problems, such as the release of oxygen from oxygen-carrying hemoglobin molecules. Diving birds may actually need more oxygen while swimming underwater than at the surface because of the additional work their muscles are doing. How do diving birds like loons overcome these physiological problems?

First, when a diving bird's face hits the water, the bird experiences a reflex, known as bradycardia, in which its heart rate drops immediately. A lower heart rate effectively lowers the metabolic demand, which means tissues (and cells) need less oxygen, a good thing on a dive. Next, blood flow to the extremities is reduced. This means that blood can go where it is most needed, such as large muscles, the brain, and the eyes. Third, some divers have more blood volume than expected for their size (e.g., darters). This means that oxygen supplies can last longer, especially if the bird is reducing blood flow to certain parts of the body. Circulating within the blood are red blood cells, which contain the protein hemoglobin. Hemoglobin is great at binding to oxygen, and the more hemoglobin an organism has, the greater its oxygen reserve, or holding tank. More blood means more hemoglobin, which means more oxygen, thereby increasing dive duration. Most vertebrates have an additional protein, called myoglobin, in their muscle, which is very good at holding oxygen. Thus, once the oxygen in the blood has fallen to very low levels, the oxygen bound to the myoglobin in the muscle comes off and enters the cytoplasm of the cell. Think of myoglobin as a holding tank for oxygen; when a loon needs more oxygen for its tissues, it opens the myoglobin valve.

Depth of dive and duration of dive are correlated. A rule of thumb is roughly 3.3 feet per second for both ascending and descending. It will take a minimum of twenty seconds for a bird to descend and ascend from 33 feet (ten seconds to descend plus ten seconds to ascend), or roughly a minute to descend to 100 feet and resurface. This does not take into account any time spent exploring on the way down or at the bottom. In penguins, both maximum duration of dives and dive depth are a function of body mass. For example, Little Penguins, with a body mass of less than 1,000 grams (about 2 pounds), rarely exceed dives of one minute or reach more than 100 feet; medium-sized penguins, of 3,000 to 6,000 grams (6 to 12 pounds), have maximum dive durations of three to six minutes and can reach 300 to 600 feet; and large penguins (e.g., King, Emperor), of 7,000 to 9,000 grams (14 to 18 pounds), can stay submerged for seven to eight minutes and reach 800 feet (or more). One female Emperor Penguin off McMurdo Sound in Antarctica was logged to a phenomenal 1,500 feet. There have been reports of loons caught in commercial fish nets set at 180 to 230 feet, which would take just over two minutes of dive time. If loons possess physiological adjustments similar to penguins and can swim just as efficiently as them, their upper limit is likely closer to 5 or 6 minutes, with a maximum dive depth of 300 feet or more. Data are limited for other loon species, but from what are available, it appears that maximal dive depth is related to body mass, just like in penguins. That is, both Common and Yellow-billed Loons can stay submerged longer than the smaller Red-throated and Pacific Loons.

How efficient are loons compared to other divers (such as grebes and auks)? To find out, we could measure how long each bird stayed underwater and then how long it stayed at the surface before diving again. If we compared the ratios of dive time/interval time among species, we could infer that between any two birds the one with the higher ratio would be the one less efficient at using oxygen. Using dive time/interval time ratios for several diving birds, one researcher concluded that loons were the second most efficient divers, followed by cormorants, then grebes, then diving ducks (e.g., canvasback). The best diver wasn't one bird but a group of them, the auks (or alcids), which include puffins, murres, and guillemots.

To be successful in its underwater niche, the loon has evolved, undergoing several anatomical and physiological modifications as well as some behavioral adjustments. The loon's major skeletal modifications include a change in the weight, shape, and length of several bones (skull, ribs, upper and lower leg, toes) and a change in the angle of attachment at several joints (lower leg). These modifications reduce drag and inefficiency by streamlining the loon's overall shape and by increasing thrust. Similarly, specific muscles are more developed (jaw, lower leg, hips) or underdeveloped (wing, breast), or the point of insertion has changed (upper leg-knee). A loon's physiological modifications are numerous and deal with temperature and salt regulation, and low oxygen levels. We have more to learn about loon physiology, and the time is ripe for an aspiring and creative ecophysiologist to seize the opportunity.

We love our loons, and we are comforted knowing that they are always there, more or less, on the same part of the lake each day. We count and observe them, paying attention to where they build a nest, when their chicks hatch, and with whom they battle or interact with throughout the breeding season. We probably know more about loon behavior than we do of any other bird in North America. Now that we know what a loon *is*—how it works and is put together—let's take a look at what a loon *does.*

Further Reading

Barr, J. F. 1973. Feeding biology of the Common Loon (*Gavia immer*) on oligotrophic lakes of the Canadian Shield. Ph.D. diss., University of Guelph, Ontario.

Clifton, G. T., and A. A. Biewener. 2018. Foot-propelled swimming kinematics and turning strategies in Common Loons. *Journal of Experimental Biology* 221: 1–11.

Clifton, G. T., J. A. Carr, and A. A. Biewener. 2018. Comparative hindlimb myology of foot-propelled swimming birds. *Journal of Anatomy* 232: 105–23.

Coues, E. 1866. The osteology of the *Colymbus torquatus*: with notes on its myology. *Boston Society Naturalist Historical Members* 1: 131–72.

Cracraft. J. 1982. Phylogenetic relationships and monophyly of loons, grebes, and hesperornithiform birds, with comments on the early history of birds. *Systematic Zoology* 31: 35–56.

Davis, L. S., and M. Renner. 2003. *Penguins.* New Haven, Conn.: Yale University Press.

Dewar, J. M. 1924. *The Bird as a Diver.* London: Witherby.

Eduardo, J., P. W. Bicudo, W. A. Buttemer, M. A. Chappell, J. T. Pearson, and C. Bech. 2010. *Ecological and Environmental Physiology of Birds.* Oxford: Oxford University Press.

Greillet, D., G. Kuntz, A. J. Woakes, C. Gilbert, J. P. Robin, Y. LeMaho, and P. J. Butler. 2005. Year-round recordings of behavioural and physiological parameters reveal the survival strategy of a poorly insulated diving endotherm during the Arctic winter. *Journal of Experimental Biology* 208: 4231–41.

Johnsguard, P. A. 1987. *Diving Birds of North America.* Lincoln: University of Nebraska Press.

Kooyman, G. L., and T. G. Kooyman. 1995. Diving behavior of Emperor Penguins nurturing chicks at Coulman Island, Antarctica. *Condor* 97: 536–49.

Kooyman, G. L., P. J. Ponganis, M. A. Castellini, E. P. Ponganis, K. V. Ponganis, P. H. Thorson, S. A. Eckert, and Y. LeMaho. 1992. Heart rates and swim speeds of Emperor Penguins diving under sea ice. *Journal of Experimental Biology* 165: 161–80.

McIntyre, J. W. 1988. *The Common Loon: Spirit of Northern Lakes.* Minneapolis: University of Minnesota Press.

Midtgård, U. 1981. The Rete tibiotarsale and arterio-venous association in the hind limb of birds: A comparative morphological study on counter-current heat exchange systems. *Acta Zoologica* 62: 67–87.

Nocera, J. J., and N. M. Burgess. 2002. Diving schedules of Common Loons. *Canadian Journal of Zoology* 80: 1643–48.

Schmid, D., D. J. H. Gremillet, and B. M. Culik. 1995. Energetics of underwater swimming in the Great Cormorant (*Phalacrocorax carbo sinensis*). *Marine Bulletin* 123: 875–81.

Schmidt-Nielsen, K. 1997. *Animal Physiology: Adaptation and Environment.* Cambridge: Cambridge University Press.

Shufeldt, R. W. 1904. On the osteology and systematic position of the Pygopodes. *American Naturalist* 38: 13–49.

Storer, R. W. 1960. Evolution in the diving birds. In *Proceedings of the 12th International Ornithological Congress,* 694–707.

Storer, R. W. 1978. Systematic notes on the loon (Gaviidae: Aves). *Breviora* 448: 1–8.

Wilcox. H. H. 1952. The pelvic musculature of the loon, *Gavia immer. American Midland Naturalist* 48: 513–73.

Williams, T. D. 1995. *The Penguins.* Oxford: Oxford University Press.

4

Pairing Up

The Behavioral Ecology of Loon Courtship

May 7, 1993. Dave Evers and I were driving around Seney National Wildlife Refuge in Michigan's Upper Peninsula, stopping at "pools" (lakes) to observe loons. Specifically, we were checking to see if the loons were banded. I met Dave in 1980 at Lake Superior State University. We lived in the same wing of Brady Hall and quickly became friends (though he was directly responsible for me intentionally missing the only class in my college career because he wanted to go birding). Little did he or I know at the time that both of us would still be conducting loon research twenty-seven years later. Dave, Joe Kaplan, myself, and several other volunteers had caught and banded more than a dozen loons on the Seney refuge. Based on our collective efforts, we knew where each loon was banded and who was paired together in the previous years. That day we confirmed that the bands on the male loon in G Pool were the same color combination as last year's male, but the female was unbanded. This was odd because last year's female *was* banded. So, unless last year's female lost all of its bands (highly unlikely), this was a new bird. This raised several questions for me. Where was last year's loon? Did she die during migration or over the winter? Where did this unbanded female come from? I was confused: I thought loons mated for life, or do they?

We continued observation until well into the night. Dave had refined the night-lighting technique for capturing loons that Judy McIntyre had attempted in the 1970s. He figured out that capture success increased if researchers targeted pairs with chicks younger than four weeks old and used a recording of a loon call (e.g., wail or chick distress) while approaching the family. The technique requires a driver, a spotter, and a netter, and equipment includes a bright spotlight attached to a portable marine battery, a large long-handled landing net, and a portable recorder. We launched when it was dark, around ten fifteen. We targeted specific loon pairs with chicks, because those pairs won't dive when approached by the boat. Having a strong drive to protect their young, the parents will not abandon their chicks and will remain on the surface, positioning themselves between the threat (us) and the chicks. If they did not have young, they would simply dive underwater and be much harder to catch. Our driver put the netter in position to catch the loon while the spotter (me) kept the light concentrated on the bird's head, specifically the eyes. We played a chick distress call from the recorder; this confuses the loon and often results in it swimming toward the boat, where it can be easily scooped up with the fishing net.

After successfully landing the loon, Dave placed two bright UV-resistant polyester colored bands on the right leg, and one colored band and one metal band on the left leg. The metal band contains an eight-digit number from the U.S. Geological Survey. This color combination allows us to separate one loon from another, though it takes some doing and patience to confirm the leg bands later using a spotting scope. Dave's refinement of the loon capture technique, along with his vision and drive, opened the door to countless new research questions. Since those fateful beginning years, more than eight thousand Common Loons have been banded and recorded, leading to the publication of more than one hundred scientific papers, theses, and dissertations.

Three days later, I sat on the shores of G Pool, keeping tabs on that same loon pair while being pleasantly distracted by migrating warblers: Chestnut-sided, Yellow, Yellow-rumped, Nashville, Black and White, Pine, Palm, and a Common Yellowthroat. I kept an eye on the loons as I bent over to check out a blooming fringed polygala (a delicate spring wildflower). Suddenly I heard a tremolo, and I looked up just in time to spy a loon flying overhead. I quickly raised my binoculars and observed bands on the legs, wondering if it was last year's female. What would the male do? Would it return and pair with last year's mate, or would he stay with this new female? If loons mated for life, which is what was generally suspected, shouldn't the male break the pair bond with this new female and reestablish the bond with its former mate? The banded female called and swam toward the new pair in G Pool. There was a tentative meeting of the three loons, some bill dipping, and shallow circle swimming, but the male remained with the new bird. The banded female was persistent and again swam at the pair, more bill dipping and circling, but the male never wavered; he remained with the new, unbanded female. The conclusion was unequivocal: loons do *not* mate for life! Here was hard evidence.

Dave and I were funded that summer by Earthwatch, a nonprofit group based in Boston that coordinates volunteers to work with scientists on their respective projects around the world. The volunteers come from across North America and countries overseas. These are highly dedicated individuals who sacrifice their time and money in the name of scientific investigation. They bring experience, knowledge, and enthusiasm to a project, but above all, they provide inspiration and hope for the future of citizen science. Earthwatch teams typically stay in group housing for one or two weeks with the principal investigator(s), assisting with meal preparation and cleaning, data collection and entry, and, in the case of our project, catching loons. Some were even lucky enough to hold a loon on their lap. My life and career have been far richer because of my interactions with these hardworking people.

That summer, we worked in four-hour shifts, gathering behavioral data on loons at the Seney refuge. Because the situation at G Pool was so riveting, we prioritized watching those loons. For the next thirty-four days, we watched a drama unfold as the banded female tried daily to break up the new pair bond. One day she took off and circled the lake a few times and then aimed her descent between the pair in an obvious effort to split them apart, but in the end, in spite of all her attempts at reuniting with the male, this never happened. On June 14, the banded female left G Pool, not to return. The male had clearly chosen a new female despite the one from last year being available. I never witnessed any physical contest between the two females. The new pair nested that season and successfully raised two chicks. The scene at G Pool poses several questions. How do loons choose mates? What qualities do they look for? Are they faithful to a lake or a mate? How long do they remain together? Welcome to behavioral ecology.

Why Behavioral Ecology?

•

Behavioral ecology bridges traditional animal behavior with the field of animal ecology. Its major tenet is that interpreting the behavior of any animal accurately is difficult unless you know its ecology, its niche, because these collectively dictate the animal's behavior. A bird of prey, such as an eagle or a falcon, will have different demands and lifestyle than a seed-eating sparrow or an insect-eating warbler, and this difference in lifestyle may influence its mating system and the level of parental involvement in incubation and chick rearing.

Consider the mating behavior of the Marsh Wren (*Cistothorus palustris*), a small brown bird that is found across North America and is commonly found in, you guessed it, marshes. Ornithologists have observed that east of the 100th meridian, Marsh Wrens practice monogamy (one male paired to one female); west of it, they practice polygamy (one male paired to more than one female). How did the difference evolve? Behavioral ecology provides the framework for understanding this observation. The 100th meridian is the imaginary line dividing the arid western United States and the humid east, which runs through the Dakotas, Nebraska, western Kansas, and Texas. To the west of the line, evaporation exceeds precipitation, whereas east of the line, precipitation exceeds evaporation. This pattern results in more standing water on the surface east of the line than west, leading to more marshes. It also results in different plant and animal communities across the two regions.

Marsh Wrens attract mates through song. Males sing a long and complex song when they establish a territory in the hope that it will attract a mate. If a male is fortunate, a female passing by will hear the lovely call and, if everything is in order, pair and mate with him. Why would a female Marsh Wren allow its mate to pair with another female in the West, but not tolerate that behavior in the East? The answer: In the arid West, marshes are in short supply, and competition among males for territories is more intense than in the East. Therefore, if a female Marsh Wren wants to gain access to good territories, her best option may be to allow her mate to practice polygamy if she wants to have any chance of mating at all. Interestingly, male Marsh Wrens in the West have more complex songs than males in the East. It seems that males with longer and more complex vocalizations get more mates. Males in the East have approximately fifty songs, while males in the West sing approximately two hundred songs. By understanding each region's climate, we are able to make sense of the wren's behavior (mating system) and the differences in song. By extension, a thorough and solid understanding of any animal's ecology will lead to a more accurate interpretation of that animal's behavior.

Behavioral ecologists interpret behavioral outcomes as a product of natural selection. In other words, observed behavioral patterns have been selected for over time because they enhance survival or reproduction; they are an optimizing agent in disguise. Behavioral ecologists expect individuals to behave in a manner that maximizes their survival (e.g., where to forage, migrate, and spend the winter) and reproduction (e.g., mate choice, number of young, amount and type of parental care) just as comparative anatomists expect the skeletomuscular system of organisms to be modified to perform tasks, like swimming or flying. Let us consider a loon's preincubation behaviors, such as philopatry (breeding-site fidelity), habitat and nest-site selection, mate choice, and intrasexual contests, and how they have been shaped by natural selection.

The Rush to Get "Home"

•

Loon aficionados in the North cannot wait for loons to return to the lakes in April and May, as soon as ice is out. Their impeccable timing each year is almost uncanny. Why are loons in such a hurry to get back to their breeding lakes each spring? Think about the following example. I enjoy playing tennis, and when I can, I go to our local courts in the evening, only to find them all occupied. So I wait (often impatiently) for someone to leave. To ensure the next time I play I don't have to wait, I get there early. Such is the case with loons returning to their lakes in the spring. Because of the competition for breeding space, it pays for a loon to get there early and stake its claim. Selection seems to favor the loons that arrive early (the early bird gets the worm). Loons do not return on the *exact* same date each year because ice-out varies with the severity of the winter, and they need open water to land. If the winter is mild, males will almost always arrive before females, suggesting that they wintered in different locations. If the winter is severe, loons will aggregate at open water, such as river mouths, large lakes, and open rivers, allowing females to catch up to their partners. Depending on latitude and severity of the winter, males may reach the lakes several days before females, or in some circumstances on the same day.

In 2011, I was fortunate to oversee a pilot study on loon migration at the Biodiversity Research Institute. If researchers had their druthers, they would ideally attach or implant a device that tracked a loon's location anywhere across the globe. Enter satellite transmitters. They have become much smaller (20 to 70 grams) and more affordable ($1,200 to $2,000) over the past couple of decades. The researcher has to choose between an external or internal transmitter. The external variety, such as a harness, fits over the bird, and most are recharged by solar energy and can last for years. They are great for eagles, falcons, and albatrosses, but they do not work on loons. We experimented with external harnesses on loons in the past, but they did not tolerate them at all, so implanting internal transmitters became the norm for loon research. This is a trickier operation than simply attaching an external transmitter to a bird: it requires a surgeon, such as a wildlife veterinarian, a portable tent, a table, and oxygen tanks (plus a myriad of smaller items). The loon's anatomy is unique, and only a few wildlife veterinarians in North America have been trained to implant satellite transmitters in loons. For this project, I hired Scott Ford, an avian veterinary specialist based in Milwaukee. I had worked with Scott on previous loon migration research and greatly appreciated his relaxed and professional manner. Each loon is anesthetized for thirty to sixty minutes while an incision is made in the body cavity to implant the transmitter. Another hole is made in the skin so the antennae can stick out. Loons are given electrolytes and a saline solution during the operation and are released when they are fully alert. For this project, we deployed six satellite transmitters in loons breeding in northern Maine.

We caught a pair of loons in July on Aziscohos Lake and discovered four months later that the female had wintered on the ocean south of Cape Cod, Massachusetts, while the male had wintered off the coast of Maine. We caught two more males and females from the same lake, though not pairs. Their transmitters revealed that the females had wintered south of Cape Cod, and the males north of it, further evidence that males and females winter in different locations. Why do the sexes winter in different locales, and is this same pattern observed in loons from different regions of the continent? There is more to investigate here, but if males consistently return to their breeding lakes across North America sooner than females, then it suggests that they may not migrate as far south as females. If this pattern holds true, it further suggests that males may have more to lose in terms of reproductive success than females if they return later. A male loon may need to return to breeding lakes sooner than a female to establish ownership and to stake a claim, as if to say, "I am here first: this is my spot, and I will defend it against any and all newcomers."

Philopatry: Returning "Home"

•

Loons are philopatric. The Greek root *phil* means "to love," as a philosopher is a lover of wisdom, and a philanthropist a lover of humanity. The Greek root *patr* means "home," as in patriot, defender of one's country or home. Therefore, a philopatric animal loves to return home. For loons, though, we have to put "home" in quotation marks because a case can be made that they spend equal amounts of time in both breeding and wintering grounds. Loon chicks are philopatric, and those hatched on lakes in northern Wisconsin, southern New Hampshire, or northwest Saskatchewan will return as adults to the area where they hatched. They will not disperse hundreds of miles or colonize new areas because they seldom wander far from where they were born. But this behavior increases the risk of inbreeding. Walter Piper, of Chapman University in California, was the first to document a case where a male loon mated with its mother. Since philopatry is the norm in loons, it appears to have been selected for, yet on the surface it is also a risky endeavor because of the complications associated with inbreeding. In humans, inbreeding among royal families in Russia and England has led to hemophilia, and among the Amish in Lancaster, Pennsylvania, to polydactyly, a condition in which newborns have an extra finger or toe. In wildlife, inbreeding causes low sperm counts, reduced fertility, and suppressed immune systems. Cheetahs, Florida panthers, black-footed ferrets, and wolves of Isle Royale in Michigan appear to be highly inbred. There is a good reason why animals disperse. Why, then, did philopatry develop and become the norm in loons?

In the first place, we can argue that if you were raised in an area successfully, that area probably met the ecological and environmental conditions needed (if it worked for you, it should work for me). This approach works for many animal species. Animals can get around the potential risks of a population becoming inbred by dispersing. Of course, success depends on the species and the initial population size, but moving a few individuals per year may be enough to maintain genetic

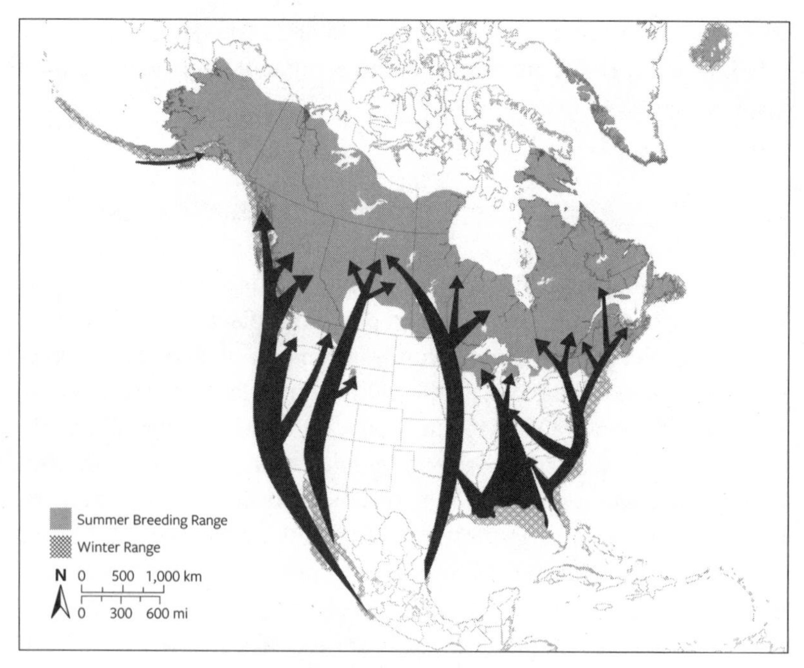

Common Loon spring migration routes. Map by Mark Burton, Biodiversity Research Institute.

diversity and prevent genetic erosion. For grizzly bears in Yellowstone National Park, for example, where the population is under five hundred animals, researchers found that bringing in one grizzly bear from outside the park every ten years for the next one hundred years could minimize inbreeding and ensure genetic diversity. Chances of inbreeding increase when populations are small. In larger populations inbreeding is largely a nonfactor because there are strong signals to avoid it. Individuals who ignore signals to avoid breeding with relatives will not be as successful as those that heed those signals. Over time, any behavior that leads to ignoring signs to avoid inbreeding will be selected against and quickly extinguished in a population. Nearly all birds avoid inbreeding by sex-biased dispersal (in birds, females more so than males, but in mammals the reverse). Loons appear to be similar in this regard; females disperse greater distances (up to 10 miles) from their natal lake than males.

Habitat Selection

•

A bird's breeding habitat must contain all the resources necessary for survival and reproduction. It must have, for example, enough food and water and suitable nesting sites. Choosing the right breeding habitat is likely the most important decision a bird will make in its lifetime. If it chooses an inferior location, one with less food, poorer nest sites, and a higher number of predators, its lifetime fitness will be adversely affected. That is to say, it may not live as long and may not produce as many young. Habitat selection is important for the majority of birds, and loons are no exception. Loons utilize freshwater lakes. One of the consequences of being a large bird like a loon is the requirement for lots of food, and generally that equates with needing lots of space to provide that food. For example, among North American woodpeckers, the largest species, the Pileated Woodpecker, is 28 inches tall and weighs between 280 and 450 grams. The smallest species, the Downy Woodpecker, is 7 inches tall and weighs from 21 to 31 grams. Not surprisingly, the Pileated occupies and requires a territory that is about eight times larger than the Downy (160 acres versus 20 acres). Another large North American bird, the Sandhill Crane, requires from 240 to 480 acres to raise a family. Loons are no different from Pileated Woodpeckers and Sandhill Cranes in this respect; they require a large lake (generally 50 or more acres in size) because larger lakes usually contain more fish than smaller ones.

More than just size of lake makes a successful breeding habitat. Anne Kuhn, a research ecologist with the U.S. Environmental Protection Agency, examined local and large-scale factors that depict loon distribution across the northeastern United States and found that loons tend to occupy lakes that are relatively large, deep, and clear. Loons also seem to prefer lakes that are close to lakes inhabited by other loons, as well as those that have a relatively high number of islands and large shoreline perimeter. With respect to physical chemistry, loons select lakes with reduced conductivity and total phosphorus, and relatively high pH. I am fairly certain that loons do not have a conductivity, pH, or phosphorus probe in their bills (or anyplace else for that matter), so they are likely selecting lakes based not on these parameters but rather on fish populations, which are indirectly or directly affected by water chemistry. Conductivity measures the dissolved salts in the water (such as chloride and magnesium ions), and most native plants and animals have evolved to function within a given range. High salt levels tend to be harmful to plants and animals that are adapted to freshwater environments. It makes sense that loons would prefer lakes with less phosphorus; too much phosphorous is the by-product of too many nutrients, which leads to algal blooms. Algal blooms negatively impact fish populations because the algae choke out plants and zooplankton, a primary food source for small fish. Lakes low in pH are more acidic, and much has been published about how acidic lakes adversely affect fish populations.

Loons use lakes that vary greatly in size. Some may be as small as 10 acres, and some as large as 500 acres. But "large" is relative. Rainy Lake on the border between Minnesota and Ontario, Canada, is the fifteenth largest lake in the world, covering some 100,000 acres, and it supports more than one hundred breeding pairs of loons. Larger lakes can hold more than one pair of loons, while lakes between 20 and 150 acres support only one pair. Lakes smaller than 20 acres are simply too small to have enough fish to support a loon family; adults nesting on these lakes typically fly to adjacent larger lakes to forage for themselves, and they use the fish in those lakes to feed their young (known as a multiple-lake territory).

This begs the question: what characteristics of lakes are most important for loons attempting to nest and raise a family? Are loons more likely to be successful raising a family on a large lake that has more than one pair, a medium lake that has only one pair, or a small lake, where adults have to feed on an adjacent lake? In 2001, Dave Evers, then at the University of Minnesota, showed that loons occupying medium-sized lakes produce more chicks per pair (0.48) than loons using large lakes (0.41) or small lakes (0.32). He discovered that from a loon's perspective not all lakes are equal. Why did loons produce more young on medium-sized lakes than the other lake types?

The problem with sharing a large lake with multiple loon pairs (known as a partial lake territory) is that the territorial boundary is undefined, and the pair has to spend time defending it from neighbors. Larger lakes attract and retain nonbreeding loons, or floaters. These individuals do not have a territory but are physiologically ready to breed and typically challenge resident loons. My studies have shown that loons occupying large lake territories are three times more likely to engage in territorial skirmishes than those occupying medium lake territories where they have the lake to themselves. They are not devoting time and energy to provisioning and guarding their young (which directly affects breeding success). Territorial skirmishes are less common in loons with a whole lake to themselves, which is probably the reason they are more successful at raising young. It is not surprising, then, that loons do poorest on small lakes (or multiple-lake territories), because they are less efficient with their time flying back and forth between lakes.

Loons have been selected to recognize and choose suitable lakes for nesting and raising a family. If a loon makes a poor lake selection, the probability increases that its genes will not be passed down to the next generation. Therefore, over time, loons that make a better decision with respect to habitat quality are favored. So, if given the choice, should a loon select a small, medium, or large lake? That is a fair enough question, but keep in mind that rarely do all three choices present themselves at once. In some regions of North America, such as in parts of Saskatchewan and Alberta, small ponds or lakes predominate, and loons there have few options; rarely will a small, medium, and large lake be found adjacent to each other. We have established that loons often return to their natal lake, or if not the natal lake, at least other lakes nearby. This suggests that if they were born on a medium-sized lake, then perhaps they have imprinted on or prefer similar-sized lakes for breeding as adults. Walter Piper has taken the lead on this question, and he found that adults tend to return and eventually settle on lakes that are similar in size and lake chemistry to their natal lake.

Territory (or lake) settlement typically has a dramatic impact on a loon's lifetime reproductive success. Sometimes loons do not get their first choice. So, if presented with either occupying a suboptimal lake or territory or not breeding at all, what should they do? Should they settle near a pair that is occupying a high-quality lake? Keep in mind that it may be difficult for a loon to acquire a territory, and an individual loon may have to wait two to five years. The foothold hypothesis suggests that individuals settle on or near high-quality territories because should one of the birds occupying the territory not return (for whatever reason), they, in theory, would be more likely to occupy it. There is evidence that some bird species adopt this strategy, but Walter Piper and his group have concluded that loons do not. He found instead that

38 percent of loons find an unoccupied territory and settle there. He also confirmed something profound: approximately 36 percent of loons evict or usurp a territorial owner of the same sex. These evictions often take the form of intense battles between the two loons.

Breeding Behavior: The Pair Bond

•

Literally all of a bird's behavior during the breeding season revolves around reproduction. Birds sing, chase other birds, and spend time with their mates because it increases their probability of mating and producing young. Darwinian natural selection has been operating on this aspect of bird behavior for hundreds of thousands of years. It is finely tuned, and the drive to reproduce is strong and the central focus of a bird's life. The life goal of any animal is to pass on its genes, and its DNA is coded in such a manner to ensure that it does. Therefore, when we ask questions about a loon's behavior during the breeding season, we have to keep this overriding theme in mind.

The great majority (about 90 percent) of all birds are monogamous, and loons are no exception. Monogamy exists when one male pairs with one female, at least for one breeding season. So, why are birds monogamous when most other animal groups are polygynous? Male mammals get away with this because female mammals can nourish the young independent of their mates through lactation. Female birds, on the other hand, cannot look after their young without paternal care. From a Darwinian perspective, sperm is cheap to produce and plentiful, whereas eggs are enormous and limited in number. This large difference in gamete size (sperms versus egg) and cost means that females have more at stake when it comes to raising young than males do, which leads to polygyny where and when possible. The fact that most mammals practice polygyny supports this idea. Female birds do not produce milk, and male birds can do just about everything a female can do to provide care, except lay eggs. Therefore, monogamy is prevalent in birds because males have the ability to assist females and contribute to their young's survival, and females alone cannot adequately provide for their young. Experimental studies have shown that if males are removed, the number of young produced decreases sharply. Loons are monogamous for a breeding season, but studies of many loon pairs across North America, including the pair in G Pool that I observed in 1993, have revealed that loons do not mate for life; rather, they practice serial monogamy, like most birds.

Intrusions and Contests

In 1994, at Manistique Lake, Michigan, I observed two loons vigorously fighting, locking bills, and powerfully slapping their wings at each other. It was intense. Both birds dove underwater, still with locked bills, and after thirty to forty seconds, they resurfaced. One bird was fleeing, the other pursuing, both flapping their wings and rowing across the lake. Eventually, the pursuer stopped chasing, and within fifteen minutes, the loon that was being pursued flew off the lake. This was a glimpse of the dark side of loon behavior that few people discuss, preferring to see loons as protectors of their young and devoted parents rather than as aggressive attackers. Yet, they are aggressive in late spring and early summer (late April until early- to mid-June) and will attack not only other loons but other waterbirds as well, such as ducks, grebes, mergansers, and cormorants.

In 2011, a wonderful colleague, Mark Pokras, director of the School of Veterinary Medicine at Tufts University, showed me a loon sternum with several holes in it. I was perplexed as to why a loon would have holes in its sternum. Mark had noticed them in both males and females. He was curious, so he placed the bill of a loon into one of the holes—it fit perfectly! Why would a loon attack another loon? The drive to reproduce dominates animals' behavior, and thus at some point in a young loon's life its motivational urge to breed overrides its fear of an aggressive encounter with a resident loon, and it tests the waters, so to speak. Because a loon that is tentative may never gain access to a territory or obtain a mate and produce young, selection has nudged young adult loons (like young bighorn sheep and bull elk) to overcome this tentativeness with aggression. Loons battle other loons for access to a territory (intraspecific aggression). But rarely does a contest between animals end in the death of one of the participants. Loons are long-lived (about thirty years). Why not live to see another day and wait its turn? Or another option is to explore and find an unoccupied lake and, with luck, a mate and start reproducing there.

Does the more experienced or the more youthful loon win the contest? Behavioral ecologists studying contests between rival males in the animal kingdom have found that typically the territorial owner wins the contest. Owners have more to lose and more motivation to defend their territory against intruders. Behavioral ecologists call this resource holding power. To use an analogy, homeowners are more likely to defend their house than picnickers their favorite picnic table they visit once a month. I might not like someone kicking me out of my favorite picnic spot, but I am not going to die fighting for it. However, were it my home, I would be more likely to defend it because I have more to lose. Territorial owners often display more aggression, are the first to attack, and will fight longer than an intruder. Breeding territorial loons are hardwired to respond with aggression to an intruding loon. They are also hardwired to attack most anything swimming in their territory (interspecific aggression), including noncompetitors, such as Mallards. One might describe a breeding loon as überaggressive. Walter Piper estimated that 16 to 33 percent of all territorial evictions in male loons are fatal for the displaced owner. In other words, one out of six loons dies during a contest, an extraordinarily high percentage compared to other species. I am unclear why this percentage is so high. One explanation is that this high fatality rate occurs when the population's age structure is skewed toward older individuals, because older individuals tend to invest more energy in reproduction and survival than younger individuals. Another hypothesis is that this high fatality rate occurs when the demand for breeding locations is intense (high competition due to overpopulation). Piper found that older males were more involved in these fatal battles than younger males, and that males fourteen years and older showed greater susceptibility to being usurped than younger males.

Researchers at Tufts University performed necropsies on 574 loons. They found that 51 percent of females and 44 percent of males had sternal punctures, and the average total number of sternal punctures for both sexes was seven. The cause of death attributed to injuries sustained during aggressive contests with other loons was statistically the same for the sexes (35 percent for females, 36 percent for males), suggesting something novel to loon researchers: females are as involved in contests as males. Researchers also found that body size was not correlated with the total number of puncture wounds. Collectively, these findings suggest that females are involved in territorial aggression and also sustain fatal injuries, just like males.

Courtship

On June 14, 1960, at Crex Meadows Wildlife Area in northern Wisconsin, William E. Southern observed what he thought was copulatory behavior in the Common Loon. He wrote up his observation and published it in *The Wilson Bulletin*. The paper described two loons chasing each other, rowing with their wings across the lake until the lead bird appeared to be tiring and the pursuer overtook and mounted it. Together they submerged, and the top bird slipped off. Today, after observing thousands of loon pairs, we know what Bill observed that day was not copulatory behavior but likely a resident loon chasing a bird without a territory. Hundreds of loon copulations have been observed and described since this written account, and they always occur on land, not in the water (I have observed about a dozen). But Bill's observation is important because it reminds us that it takes time to get to know animals well enough to be able to interpret their behavior correctly. We cannot understand loon behavior (or any animal's, for that matter) by observing them for a single day. Often it takes weeks, if not months, and in many cases years. The caveat? Spring cannot be defined by one warm or one cold blustery day, nor can we make inferences after observing an animal for a few hours.

Loon courtship consists of synchronous facing away, bill dipping, circle swimming, and perhaps even some jerk diving, but nothing as overly elaborate or outlandish as found in other bird species. The male does not go overboard, nor does the female appear to be testing the male in any overt way. It is more of a greeting and reuniting ceremony. The pair bond is further strengthened by doing everything together, such as preening, foraging, and patrolling their territory. Courtship behaviors are ritualized, or stereotyped. They are performed in a certain manner and sequence, and they are clear and unambiguous. Think about it: every loon knows how to behave toward a member of the opposite sex; if they do not face away, or bill dip, or circle swim in the right manner, it will result in unsuccessful mating. But where do courtship behaviors come from? How did they evolve and become fixed in the DNA of an animal?

At a basic level, they evolved from nonsignals. The courtship behaviors of loons were already being performed by them in everyday life. When two strange birds meet, one (or both) may face away to avoid looking directly into the eyes of the other (humans do this as well). By looking away they are sending a signal that says, "I am not a threat," thereby reducing aggression. All loons dip their bills into the water to wet them prior to sticking them into their feathers to preen. And circle swimming was originally part of diving, a foraging behavior. Therefore, loon courtship behaviors observed today all evolved from nonsignals that they performed daily. Other birds, such as gulls and ducks, incorporate many of the same behaviors into their courtship rituals because these were also part of their everyday movements.

Once territories and pairing are settled and established, and successful courtship is achieved, loons begin the next phase of the reproductive cycle: nesting. The nesting period is a critical time for any bird because the choices made (such as where to build the nest, and if and when to abandon it) are critical for increasing the probability of successfully hatching chicks.

Further Reading

Alcock, J. 2012. *Animal Behavior: An Evolutionary Approach.* 10th ed. Oxford: Oxford University Press.

Alvo, R. 2009. Common Loon, *Gavia immer,* breeding success in relation to lake pH and lake size over 25 years. *Canadian-Field Naturalist* 123 (2): 146–56.

Evers, D. C. 1993. A replicable capture method for adult and juvenile Common Loons on their nesting lakes. In *Proceedings from the 1992 Conference on the Loon and Its Ecosystem: Status, Management, and Environmental Concerns,* 214–20. Bar Harbor, Maine.

Evers, D. C. 2001. Common Loon population studies: Continental mercury patterns and breeding territory philopatry. Ph.D. diss., University of Minnesota, St. Paul.

Gill, F. B. 2007. *Ornithology.* 3rd ed. New York: W. H. Freeman.

Gingras, B. A., and C. A. Paszkowski. 1999. Breeding patterns of Common Loons on lakes with three different fish assemblages in north-central Alberta. *Canadian Journal of Zoology* 77: 600–609.

Gray, C. E., J. D. Paruk, C. R. DeSorbo, L. J. Savoy, D. E. Yates, M. D. Chickering, R. B. Gray, et al. 2014. Body mass in Common Loons (*Gavia immer*) strongly associated with migration distance. *Waterbirds* 37: 53–63.

Kroodsma, D. E., and J. Verner. 2020. Marsh Wren (*Cistothorus palustris*), version 1.0. In *Birds of the World,* edited by A. F. Poole. Ithaca, N.Y.: Cornell Lab of Ornithology. https://birdsoftheworld.org/bow/species/marwre/cur/introduction.

Kuhn, A., J. Copeland, J. Cooley, H. Vogel, K. Taylor, D. Nacci, and P. August. 2011. Modeling habitat associations for the Common Loon (*Gavia immer*) at multiple scales in northeastern North America. *Avian Conservation and Ecology* 6 (1): 4.

Paruk, J. D. 1999. Behavioral ecology in breeding Common Loons (*Gavia immer*): Cooperation and compensation. Ph.D. diss., Idaho State University, Pocatello.

Piper, W. H., K. M. Brunk, G. L. Jukkala, E. A. Andrews, S. R. Yund, and N. G. Gould. 2018. Aging male loons make a terminal investment in territory defense. *Behavioral Ecology and Sociobiology* 72: 95.

Piper, W. H., D. C. Evers, M. W. Meyer, K. B. Tischler, J. D. Kaplan, and R. C. Fleischer. 1997. Genetic monogamy in the Common Loon (*Gavia immer*). *Behavioral Ecology and Sociobiology* 41: 25–31.

Piper, W. H., J. S. Grear, and M. W. Meyer. 2012. Juvenile survival in Common Loons *Gavia immer:* Effects of natal lake size and pH. *Journal of Avian Biology* 43: 280–88.

Piper, W. H., J. N. Mager, C. Walcott, L. Furey, N. Banfield, A. Reinke, F. Spilker, and J. A. Flory. 2015. Territory settlement in Common Loons: No footholds but age and assessment are important. *Animal Behaviour* 104: 155–63.

Piper, W. H., M. W. Palmer, N. Banfield, and M. W. Meyer. 2013. Can settlement in natal-like habitat explain maladaptive habitat selection? *Proceedings of the Royal Society.* DOI:10.1098/rspb.2013.0979.

Piper, W. H., J. D. Paruk, D. C. Evers, M. W. Meyer, K. B. Tischler, M. Klich, and J. J. Hartigan. 1997. Local movements of color-marked Common Loons. *Journal of Wildlife Management* 61: 1253–61.

Piper, W. H., K. B. Tischler, and A. Dolsen. 2001. Mother-son pair formation in Common Loons. *Wilson Bulletin* 113: 438–41.

Piper, W. H., K. B. Tischler, and M. Klich. 2000. Territory acquisition in loons: The importance of take-over. *Animal Behaviour* 59: 385–94.

Piper, W. H., C. Walcott, and J. N. Mager. 2008. Fatal battles in Common Loons: A preliminary analysis. *Animal Behaviour* 75: 1109–15.

Piper, W. H., C. Walcott, J. N. Mager, and F. J. Spilker. 2008. Nestsite selection by male loons leads to sex-biased site familiarity. *Journal of Animal Ecology* 77: 205–10.

Reece, J. B., L. A. Urry, M. L. Cain, S. A. Wasserman, P. V. Minorsky, and R. B. Jackson. 2014. *Campbell Biology.* 10th ed. Boston: Pearson Education.

Southern, William E. 1961. The courtship behavior of the Common Loon. *Wilson Bulletin* 73: 280.

Winkler, D. W., J. Shamoun-Baranes, and T. Piersma. 2016. Avian migration and dispersal. In *The Cornell Lab of Ornithology Handbook of Bird Biology,* edited by I. J. Lovette and J. W. Kirkpatrick. West Sussex, UK: John Wiley and Sons.

5

Taking Turns

NESTING BEHAVIOR AND ECOLOGICAL TRADE-OFFS

May 24, 1995. Turtle Flambeau Flowage, northern Wisconsin. I tied my very old wooden boat passed down from my grandfather to a birch tree and walked up and over a raised mound via a footpath, which led me to my field assistant, Lucy Vlietstra. My research that summer was quantifying sex roles during incubation, specifically, who sits longer on the nest, and does that vary at all during the twenty-eight-day incubation period? I also wanted to find out how long loons incubated their eggs before exchanging with their mates and was curious to uncover any patterns in nest exchange between loons using different-sized lakes. Lucy was recording data on a loon nest using a spotting scope and binoculars from roughly 75 yards away. Her four-hour shift was done, and it was my turn to replace her. She pointed out a Bald Eagle that had landed in a white pine about forty minutes earlier, roughly 30 to 40 yards from the loon nest. I could see the male loon on the nest, alert but not anxious. I looked at the eagle: it seemed content perching. Lucy decided to wait with me for a few more minutes, and we sat in silence and just watched. Suddenly, the eagle swooped down off its perch heading directly for the loon. Before it could fully slip into the water, the eagle sunk its talons into the loon's back. The nest sat on the edge of a floating bog, with one side facing a steep 3-to-4-foot incline over the water, while the other side faced aquatic vegetation in about a foot of water. This location turned out to be fortunate for the loon, as the deep water gave it space to writhe and attempt to escape.

The struggle continued for twenty seconds, but eventually the loon freed itself and rowed across the surface of the lake, calling wildly. The eagle remained on the edge of the nest for a full two minutes, before leaping into the air, sweeping its massive wings. After the male left, an hour passed before the female returned, shuffled up the nest, turned the eggs, and resumed incubating. During the interim, no predator had noticed the exposed eggs (the pair was fortunate). The male returned to the nest the following day, and the pair resumed their normal incubating roles. Two chicks successfully hatched from the nest on June 11.

In an odd sort of way, working on a summer-long field research project brings peace and comfort despite the sixteen-to-eighteen-hour workdays, especially if working in an idyllic setting with great people. In 1995, Dave Evers and I switched our study site from Seney National Wildlife Refuge to the Turtle Flambeau Flowage, a large reservoir, roughly 13,000 acres, home to nesting eagles, Ospreys, and more than twenty-five Common Loon pairs. Most of the shoreline was undeveloped, and many places had no sign of human activity. Loons were everywhere, and it often felt like we were in Canada. My lead field assistant, Lucy, was dependable, hardworking, enthusiastic, and always positive. She was at ease handling the birds and working with people. Most of all, I appreciated her "do whatever it takes to get the research done" attitude. We lived on peanut butter and jelly sandwiches, got caught in thunderstorms, baked in the hot sun, and endured hordes of mosquitoes, but always a resolute smile on her face greeted me.

Then there was Jeff Wilson, a Wisconsin Department of Natural Resources biologist, who greeted me with open arms. Jeff is an extrovert and enthusiastic and knowledgeable about wildlife. He was full of ideas, helped by showing me the reservoir and the general locations of the loon nests, and offered tips on how to avoid the underwater stumps when navigating through the narrows. He saved me hours of time and potentially numerous broken cotter pins, but most important he gave me his friendship. He is a world-class individual and has more stories than anyone I have ever met. Should you ever be blessed one day to sit around a campfire with Jeff, you will be in for a real treat. With such inspirational people around me, the field season that summer was highly memorable.

Nest Site Selection

•

Under the adaptationist model, nest site selection by loons, and by other birds, should be under the influence of natural selection. It is a key choice made in the life history of a breeding animal that may make the difference between a successful or failed nest. As such, it directly relates to reproductive success. The location of a nest can determine how much exposure to the elements it receives (abiotic factors, such as rain, wind, sun). In addition, its location can have a direct effect on competition from other members of the same species, and on predation (both examples of biotic factors). In many birds, predation is a major evolutionary factor determining the location of nest sites.

Anyone who has spent time searching for loon nests quickly discovers they prefer islands to mainland shorelines, sheltered areas to exposed, and, if possible, a location near a steep versus gradual slope. Should you be faced with locating a loon nest, here is my advice: determine the direction of the prevailing winds, investigate the lee side of any island or sheltered bay or cove, and look for places with a steep drop-off. The fact that loons prefer islands to mainland shorelines suggests that natural selection is operating on this feature. Paul Radomski, of the Minnesota Department of Natural Resources, analyzed data from thirty-five lakes and 258 loon nests and determined that loons were 4.2 times more likely to nest on an island than a mainland shoreline. Why islands? Raccoon predation is a major factor in nest failure. Several studies have shown that more than half of the loon nests in a given area fall to this type of predation. In a Wisconsin study, roughly five hundred nests were subject to mammalian predation because they were located on the mainland shoreline. By nesting on an island, loons greatly increase their probability of nesting successfully. Yes, avian predators, such as gulls, crows, and ravens, may also depredate an unattended nest, but collectively they exert less selection pressure than raccoons on loon nesting success.

Radomski found that loons nesting on a large lake choose the side of an island or a shore with the least amount of fetch. Winds across an unobstructed expanse of water generate large waves that can ride up the lip of the loon nest, flooding and inundating it, and causing it to be abandoned. This is why few loon nests are found facing the prevailing wind direction. Why do loons nest near a steep drop-off? Our observation of the eagle at Turtle Flambeau Flowage supports the hypothesis that edge nesting can aid loons in escaping predators. We concluded that if that nest had lacked the steep drop-off, the loon would likely have been killed by the eagle.

Does the male or the female, or do both choose the nest-site location, and does it vary from year to year or remain mostly the same?

Again under the adaptationist approach, if familiarity with a site confers reproductive advantages, and the site is then successful, the best strategy should be to use the site for as long as it is successful (if it ain't broke, don't fix it) and switch if it is unsuccessful. This behavioral pattern is known as the "win-stay, lose-switch" rule, and many animals, especially birds, employ it. Therefore, if a loon nest is successful, they should keep using it from year to year, but should it fail, they should switch to a different nest location. When a nest fails, it is probably because of mammalian predators like raccoons and skunks. The win-stay, lose-switch rule appears to have evolved in loons in response to mammalian nest predation. This rule explains why so many loon nests are in the same spot each year and also why some are in a different location the following year.

Optimal Clutch Size

•

Once a pair of loons has established a suitable nest site, what is the optimal number of eggs, or clutch size, the female should lay? Imagine a scenario with three distinct loon populations: A, B, and C. Females in population A lay one egg only; in Population B they lay two eggs; and in Population C, three eggs. Further imagine that predation and mortality are equal among the populations: they each experience the same risks from nest predation and chick survival and have the same survivorship from year to year. Run the following experiment for fifty years: mark all the young that are produced from each population, follow them to adulthood, and keep a tally of the total number of fledged young (and of those young, how many breed and reproduce young of their own).

Behavioral ecologists have studied such situations in many different species of birds (and other vertebrates and invertebrates). A bird can lay more eggs, but the offspring of those eggs will not necessarily reach adulthood and reproduce. In other words, a trade-off exists between the number of eggs laid and the fitness of the offspring. The idea is that parents can provide only so much care for their young, and if the number of young produced is increased, it will usually come at the expense of the fitness of the young. Laying additional eggs and hatching more young may actually depress the survivability of other members in the clutch because the parents cannot provide enough food for all of them. David Lack, a British ornithologist, was the first to propose that an intermediate clutch size may produce the greatest number of survivors. He further proposed that each bird species would evolve the most productive clutch size for its environment and niche. However, larger brood sizes not only increase pressures on the young but also on the parents, who must spend more time gathering food to feed the larger brood, time that could be spent on themselves (e.g., to fatten up for migration). This is the second trade-off that pops up in any discussion of optimal clutch size. If a parent puts too much energy into current reproduction, it may compromise future reproduction. Through experimental manipulation of clutch size (subtracting and adding eggs), researchers found that clutch size generally equals the most productive brood size for many bird species but not all. Both trade-offs explain current optimal clutch size in birds. If you add eggs to a clutch, chicks experience lower survival rates, and parents tend to have lower reproductive output the following year.

The trade-offs do not operate at the same level for all bird species. For example, a Northern Cardinal that puts too much energy into rearing offspring one year may depress its condition such that the following year she does not lay the same number of eggs. The math looks like this: in the first year Cardinal A rears six young, in the second it rears two young, and in the third it rears three young, for a total of eleven young. Cardinal B rears four young in each of the first, second, and third years, for a total of twelve young. In this scenario, Cardinal B will have contributed more genes to the next generation than Cardinal A. Population modeling shows that Cardinal A will be outcompeted in the long term by Cardinal B, the bird that did not overexpend its current reproduction. How many eggs should Common Loons lay? As Lack predicted, natural selection settles on the clutch size that maximizes reproductive success for each species, given that clutch size will almost always vary depending on external factors, such as weather and prey density. The clutch size of loons is predominately two eggs, but one or three egg clutches are found on occasion.

The influence of natural selection is evident when we compare clutch sizes of similar-sized groups of birds in the tropics and temperate regions. Birds in the tropics lay fewer eggs than birds in the temperate region because they have more than one brood per season. But why would they lay fewer eggs per clutch? Given the intense egg predation by snakes and mammals that tropical biologists have observed, why put all your eggs in one basket? The better strategy may be to lay fewer eggs but have more broods, with the hope that one brood will be successful. Natural selection is also at work in some species of birds when food is plentiful. For example, when vole densities are high, one of their major predators, the Snowy Owl, will lay seven to eleven eggs, and when vole numbers are low, they will lay only three to four eggs. Natural selection is also evident when we compare the clutch size of ground-nesting birds and cup (tree) nesters. Ground nesters, which are subjected to higher predation rates, lay more eggs, from six to eighteen, than cup nesters, whose average clutch size is four to five eggs.

Eggs are costly for females to produce. Every day that a chicken lays an egg, about 1.8 percent of her body weight goes into that egg. For comparison, the human female deposits about 0.019 percent of her body weight into a developing fetus and placenta. Egg production for ducks and geese requires 52 to 70 percent of daily energy intake (considerably less for smaller birds, like finches and sparrows). The incubation period may increase basal metabolic rates in birds by a factor of four or five. That would be analogous to a person with an average intake of 2,000 calories per day jumping to 8,000 (or more).

Loon eggs are large; they weigh between 140 and 160 grams and are roughly three times the weight of a chicken egg. As a rule, the larger the bird, the smaller its egg is in relation to its body size. Ostrich eggs are huge but comprise only 1.7 percent of the bird's body weight. Wrens lay tiny eggs, but they can be up to 13 percent of their body weight. Birds that hatch naked, featherless young lay proportionally smaller eggs than birds that hatch fully feathered young. For example, the eggs of warblers, sparrows, and blackbirds, which hatch altricial young (naked), are 4.5 to 7.7 percent of their respective female's weight. In contrast, eggs from owls, waterfowl, and shorebirds, which hatch precocial young (fully feathered), weigh 11.5 to 21.1 percent of their respective female's weight. Two loon eggs will weigh around 8 to 10 percent of the female's body weight, which is on the low end of species that produce precocial young. A loon chick is subprecocial in that the young are fully feathered but dependent on parents for food. This is different from a recently hatched grouse, which can catch or find its own food soon after birth. Though the cost of producing a loon egg does not seem extraordinary by any means when compared to other birds, these numbers do reinforce the notion that the cost of producing an egg is substantial for most avian females.

It is not surprising, then, that in preparation for the exacting strains of egg laying, many female birds increase their body weight by 20 to 25 percent. Loons are long-lived, and females will lay eggs for fifteen to twenty years or more. Because of the cost to a loon that lays three eggs and the costs associated in feeding the young (which is probably the more limiting factor), the bird will enter the fall-winter cycle depleted, possibly severely so. In other words, her bank account will be in the red, and she may find herself in the next breeding season incapable of laying more than one egg, if she can lay any at all. During one nesting season, I observed loons for eighteen hours a day and found that males spent 40.5 percent of their time each day feeding during the entire twenty-eight-to-twenty-nine-day incubation phase, whereas females spent 46 percent of their time feeding during the first two weeks of incubating before tapering down to 40 percent in the last two weeks. This finding suggests that during incubation females are recovering from costs associated with egg laying. Should a clutch fail for some reason, a following clutch is laid anywhere from eight to eighteen days later, likely the amount of time needed for the female to recover from the initial egg production. If a nest fails within the first eighteen days after the initial egg is laid, loons are almost certain to renest. But if the nest fails after eighteen days, then the odds of renesting diminish. I once studied a pair of loons that had a long history together (eight years), when their nest failed twenty-two days into incubation. This turned out to be destabilizing, because the pair broke up and the female paired with another male the following year. It is important to note that if loons do not raise young one year, it is not the end of the world, given that a female will have fifteen years or more to produce an offspring (or two) to replace herself in the population when she dies.

Incubation Duties as Assigned

•

Part of my dissertation research investigated sex roles of loons during the breeding season. For the nesting period, I observed the pairs' patterns of nest exchange, length of incubation bouts, and overall time incubating the eggs. Over a four-year period, my Earthwatch crews and

I monitored eleven loon pairs continuously from a half hour before sunrise to a half hour after sunset. We logged 1,653 observation hours over ninety-nine days (each pair was sampled nine days during the incubation period). We found that there was no significant difference between the sexes in the total time (in minutes) spent incubating the eggs. Males shared incubation duties equally with females. In some cases, females sat on the nest more than their mates, and in other pairs we observed the opposite. Why did some males spend more total time incubating than their partners did? Was it because they were younger and paired with more experienced females who took advantage of their willingness to do more than their fair share? Individual and pair variation are the norm in loons: not all males, or all females, or all pairs behave a certain way with respect to nest attentiveness. Or perhaps territory type was a factor? Since partial lake territories have more intrusions than whole lake territories, perhaps males are busier patrolling on the former than the latter and consequently spend less time on the nest. Answering one question often leads to more questions, and when that happens, there is a saying in our profession: that's when you are doing good science.

To see if there were any changes in nest attentiveness of pair members between years, I studied two pairs. Each pair had nested successfully the year before, and each had paired with the same mate the following year. I was not sure what to expect, but each pair behaved as they had the previous year. In pair A, nest attentiveness in year one was 53 percent female and 39 percent male, and in year two was 51 percent female and 42 percent male. In pair B, nest attentiveness in year one was 53 percent female and 46 percent male, and in year two 52 percent female and 47 percent male. I had more questions. Will this pattern hold up if a larger sample is obtained? Does parental effort between pair members stabilize and remain unchanged over time? Hypothetically, say a female sits on the nest for more time than her partner does, and does so for five consecutive years. Will there ever be a time when her partner will sit for more time than she does? Are the roles fixed once the pair become established? Many of us are familiar with long-term partnerships. Who does the majority of the cooking, pays the bills, mows the lawn? Did those roles change after the first year of partnership? If not, you may see what I am getting at: patterns of behavior may be hard to change once they become established.

Incubation Length

•

How long do loons incubate before being replaced by their mates? One pattern I found was that loons sat for shorter times in the beginning of the incubation cycle (the first nine days) and increased the amount of time they incubated as the cycle progressed. The trend showed that both sexes incubate for 2 to 4 hours in the first two weeks, and 3 to 6 hours in the latter two weeks prior to hatching. The average incubation bout was 4.5 hours, which is double what was reported in an earlier study. Two of the eleven pairs nested on a small lake but foraged on a larger adjacent lake. I observed that pair members would sit for extraordinary periods while the partner was away. In pair A, the average incubation bout was 6 hours, and in pair B, 7.5 hours. The longest time between nest exchanges was 13 hours 54 minutes. This shows that incubation bouts in loons are not fixed, and loons can adjust their schedules to adjust to the amount of prey available in a territory.

We know there is great diversity in incubation patterns among birds. How does the incubation bout length in loons compare to other similar species? If sexes look alike, then they both take equal turns in incubating the eggs, and if the plumages are dissimilar, the drab, less conspicuous partner does the incubating (such as a mallard). But nature loves exceptions to its rules. For example, the brightly colored male Rose-breasted Grosbeak incubates the eggs just as long as its less conspicuous mate. Cormorants' incubation bouts are one to three hours, while male sandpipers and plovers incubate during daytime hours, and females at night.

The purpose of incubating eggs is to keep the developing embryo at a temperature that permits its growth. Too hot or too cold and the embryo will die. The optimal temperature for growth and development is 37° to 38°C, and birds do their best to keep their eggs at this temperature. That is why the division of incubation duties between parents is so patterned and predictable: the internal temperatures of eggs remain remarkably stable because of this partnership and cooperation. But if one bird does not carry its weight, the other member may simply abort altogether. This was the case with one pair of loons I observed: the male was simply not contributing its share. The female, after being the predominant incubator for a week, simply got off the nest. Crows got the eggs. Later, it was found this male had elevated levels of mercury.

Mercury toxicity is not uncommon in loons (see chapter 11), and one of its symptoms is a lack of attentiveness. Serious problems occur if the embryos are exposed to temperatures below 35°C or above 40.5°C. This is the reason why nest attentiveness in birds is generally above 95 percent and often closer to 99 percent. Most birds provide heat to the developing embryo with a brood patch, a localized area of the abdomen where the bird loses its feathers. During incubation, this area becomes highly vascularized, and the open blood vessels transfer more heat to the embryo. Loons (like penguins and a few other bird species) lack a brood patch altogether, which is why good parental cooperation is essential to the survival of their eggs.

Patterns of Nest Exchange

•

When it comes to nest exchanges, some pairs of birds maintain a very regular schedule, and others do not. My data suggest that loons nesting on whole lake (medium-sized) territories have more coordinated schedules than loons nesting on partial lake (large-sized) territories. For example, 80 percent of the first nest exchanges of the day on whole lake territories took place within 1.5 hours of sunrise, compared to only 20 percent for loon pairs nesting on large lakes. Over the first nine days of incubation, one pair that I observed on a medium-sized lake had the following times for the first nest exchange of the day: 636, 613, 614, 550, 552, 610, 639, 622, and 616 a.m. (remarkably tight and consistent), while a different pair on a large-sized lake had the following times: 648, 538, 703, 705, 510, 554, 455, 830, and 848 a.m. (more variable and less consistent). Although patterns of nest exchange vary quite a bit, territory type may influence how tight or in sync patterns of nest exchange are.

Loons on large lakes had shorter incubation bouts (on average, thirty-four minutes shorter) and lower overall nest attentiveness (95 percent compared to 97 percent) than loons on medium-sized lakes. One major difference between the two lake sizes is that loons on large lakes spent twice as much time interacting with other loons (mostly intruders) while incubating than loons on medium-sized lakes. Being next to neighbors has a cost for loons; additional energy is spent defending and patrolling their territory. Some pairs on large lakes have neighbors on both sides of their territory, which means they have to defend their borders twice as much as loons that have just one border to defend. Larger bodies of water also attract nonbreeding loons, or floaters. An average loon population includes 15 to 45 percent floaters. These unpaired loons are trying to acquire a territory. They will challenge resident loons on large lakes more often than on medium-sized lakes because, logistically, they are closer together and it is easier to do. During my research, it was common for a nesting loon (often a male) on a large lake to leave the nest and approach an intruder. But this left the eggs uncovered, which resulted in lower nest attentiveness and lower nesting success (the number of young fledged per nesting pair). Avian predators are good at finding and remembering the location of bird nests and equally as good at cracking and eating eggs. Loons on large lakes also spent more time interacting with intruder loons in the evening and nighttime compared to those on medium-sized lakes, which were rarely visited by other loons at times other than early morning. There was also a tendency for females on large lakes to sit on the nest at night more often than females on medium-sized lakes (59 percent versus 47 percent). Males on large lakes with neighboring loons are likely patrolling their territories during the late evening and continuing through the night.

In addition to facing the challenge of heat regulation, incubating loons often endure small, biting blackflies that draw blood. One species, *Simulium annulus,* is particularly attracted to loons. Female blackflies likely cue in on oil from the loon's uropygial gland and will fly across open water, if necessary, to obtain a meal. Males have rudimentary mouthparts and do not bite but instead feed on nectar or plant sap; some even pollinate blueberries. Upon obtaining a meal, fertilized females use the nourishment to lay eggs, which they deposit in fast-flowing streams and rivers, some as far as 10 miles away. From there, the eggs develop slowly over the remaining summer months and overwinter as pupae. They hatch the following spring, emerging when the ice has melted and temperatures have warmed (above 50°F). They live another two to three weeks, during which time securing a meal is paramount to their survival. Because of the annoying challenge they pose, it is not uncommon during years with high blackfly numbers to see increased nest abandonment by loons.

The concept of trade-offs is important in any discussion of an animal's life history characteristics and traits. A certain trait can be increased but only at the expense of another. A bird can lay additional eggs and raise more young, but that means it will have less energy for its own daily needs. The bottom line is that surviving is hard—for loons, other birds, other animals. Few animals have the luxury of not making trade-offs at some point in their life history. There are always limitations: only so much time in a day or night, only so much food or suitable nesting sites, only so many mates or so much territory. Trade-offs involving egg size, number of eggs, and an animal's investment in time or energy abound everywhere in nature. Trade-offs also occur when birds communicate, and loons in particular have some of the most iconic calls in the entire bird world.

Further Reading

Alcock, J. 2012. *Animal Behavior: An Evolutionary Approach.* 10th ed. Oxford: Oxford University Press.

Clutton-Brock, T. H. 1991. *The Evolution of Parental Care.* Princeton, N.J.: Princeton University Press.

Gill, F. B. 2007. *Ornithology,* 3rd ed. New York: W. H. Freeman.

Gowaty, P. A., and D. W. Mock. 1985. *Avian Monogamy.* Ornithological Monographs. No. 37. Washington, D.C.: American Ornithologists' Union.

Koenig, W. D. 2016. Ecology of bird populations. In *The Cornell Lab of Ornithology Handbook of Bird Biology,* edited by I. J. Lovette and J. W. Kirkpatrick. West Sussex, UK: John Wiley and Sons.

McCann, N., D. Haskell, and M. W. Meyer. 2004. Capturing Common Loon nest predators on 35mm film. *The Passenger Pigeon* 66: 351–61.

McIntyre, J. W. 1988. *The Common Loon: Spirit of Northern Lakes.* Minneapolis: University of Minnesota Press.

Paruk, J. D. 1999. Behavioral ecology in breeding Common Loons (*Gavia immer*): Cooperation and compensation. Ph.D. diss., Idaho State University, Pocatello.

Paruk, J. D. 2000. Incubating roles and patterns in Common Loons: Old history and new findings. *Proceedings of a Symposium from the 1997 Meeting,* edited by J. McIntyre and D. Evers. American Ornithologists' Union. North American Loon Fund. Holderness, N.H.

Piper, W. H., D. C. Evers, M. W. Meyer, K. B. Tischler, J. D. Kaplan, and R. C. Fleischer. 1997. Genetic monogamy in the Common Loon (*Gavia immer*). *Behavioral Ecology and Sociobiology* 41: 25–31.

Piper, W. H., J. S. Grear, and M. W. Meyer. 2012. Juvenile survival in Common Loons *Gavia immer:* Effects of natal lake size and pH. *Journal of Avian Biology* 43: 280–88.

Piper, W. H., J. N. Mager, C. Walcott, L. Furey, N. Banfield, A. Reinke, F. Spilker, and J. A. Flory. 2015. Territory settlement in Common Loons: No footholds but age and assessment are important. *Animal Behaviour* 104: 155–63.

Piper, W. H., M. W. Palmer, N. Banfield, and M. W. Meyer. 2013. Can settlement in natal-like habitat explain maladaptive habitat selection? *Proceedings of the Royal Society.* DOI:10.1098/rspb.2013.0979.

Piper, W. H., J. D. Paruk, D. C. Evers, M. W. Meyer, K. B. Tischler, M. Klich, and J. J. Hartigan. 1997. Local movements of color-marked Common Loons. *Journal of Wildlife Management* 61: 1253–61.

Piper, W. H., K. B. Tischler, and A. Dolsen. 2001. Mother-son pair formation in Common Loons. *Wilson Bulletin* 113: 438–41.

Piper, W. H., K. B. Tischler, and M. Klich. 2000. Territory acquisition in loons: The importance of take-over. *Animal Behaviour* 59: 385–94.

Piper, W. H., K. B. Tischler, and A. Reinke. 2018. Common Loons respond adaptively to a black fly that reduces nesting success. *The Auk* 135: 788–97.

Piper, W. H., C. Walcott, and J. N. Mager. 2008. Fatal battles in Common Loons: A preliminary analysis. *Animal Behaviour* 75: 1109–15.

Piper, W. H., C. Walcott, J. N. Mager, and F. J. Spilker. 2008. Nestsite selection by male loons leads to sex-biased site familiarity. *Journal of Animal Ecology* 77: 205–10.

Radomski, P. J., K. Carlson, and K. Woizeschke. 2014. Common Loon (*Gavia immer*) nesting habitat models for north-central Minnesota Lakes. *Waterbirds* 37: 102–17.

Vlietstra, L., and J. D. Paruk. 1997. Predation attempts on incubating Common Loons, *Gavia immer,* and the significance of shoreline nesting. *Canadian Field Naturalist* 111: 656–57.

6

Wails, Yodels, and Tremolos

The Call of the Loon

Few sounds capture the attention of lakeshore visitors more than the call of a loon. We all have a memory of when and where we heard it first. For me, it was late afternoon, the second week of August 1979, on Chickenbone Lake, Isle Royale National Park. I had just set up my tent when I heard these unusual sounds coming from the far side of the lake. I could not identify them or the birds producing them. Another camper told me they were loons, and they were making the tremolo call. I was raised in a city, far from breeding loons, and this was my first time backpacking and camping away from home. Such experiences were foreign to me. Later that night, I would hear the loons make their famous wail call. Many of us hear loons long before we see them. These vocalizations are an integral part of loon communication with their mates, rivals, and other threats, such as humans kayaking too close to their nests.

In 1993, Charlie Walcott came to Seney National Wildlife Refuge to record loon vocalizations. At that time, Charlie was a professor of neurosciences at Cornell University and director of the Cornell Lab of Ornithology. Dave Evers invited him to stay with me and asked if I could show him around the refuge. We needed a large place that could house us along with six to eight volunteers, and we were lucky to find a bed-and-breakfast in Germfask, about 10 miles down the road from Seney. The house had numerous rooms, a large kitchen, and a spacious meeting room, ideal for our Earthwatch teams. It was also fairly luxuriant, with plush carpeting and nice furniture, atypical for a field research station, which is usually scaled down to accommodate dirty boots, wet socks, ticks, and loon poop (which often covered one's pants). I was a bit uncomfortable with the setup because for the past decade I had lived in small cabins with used furniture and questionable carpeting—some lacking even electricity and water—so it was awkward for me to call this extravagant B and B my home for the next five months.

I moved in on May 1, and Charlie arrived shortly afterward, just as the loons were returning. He was exploring new research ideas, and this pilot study was to vocal tag (record) the yodel of banded male loons for a couple of years to see whether they keep the same vocal signature from year to year. Scientists were uncertain about whether a loon's call could change over its lifetime. That spring, Charlie and I shared meals together. I live for conversation, and Charlie and I hit it off right away. We talked about a great many things, mostly loons but also politics, sustainable economics, renewable energy, and science education. One morning he walked downstairs at 4:30 a.m. to find me already up and reading a college textbook, *Animal Behavior.* (I picked it up at a used bookstore, and it was the best two-dollar investment of my life.) Charlie was curious about my motivation for reading the book. I told him that I simply loved learning and that the book grabbed my interest. By the end of the summer I had read the book in its entirety, pored over its numerous examples, and annotated every page.

Before Charlie left for home, he pulled me aside and suggested that I think about returning to higher education to get a doctorate. That time with Charlie changed the course of my life: our conversations really got me thinking about pursuing my doctorate. I visited Charlie at Cornell, and in between seminars he introduced me to professors and graduate students. I toured five other college campuses that fall, and although I did not attend Cornell, I appreciated Charlie's interest in me and my development as a scientist. As I got to know him more over the years, I found that he is an individual of great humility, a passionate science educator and advocate, a person of impeccable character and a magnanimous heart. I can't think of a more noble combination. I try to do for others what Charlie did for me.

Why Do Birds Vocalize?

•

Birds are vocal, in part, because they have wings and can fly away from potential predators. In essence, they can afford to make noise because they can get away with it. Wings have given birds certain kinds of freedoms that other animals do not enjoy. Loon calls are especially distinctive and easily recognized. Each one has a specific meaning that allows the pair members to communicate with each other. How did those different calls evolve, and what factors or forces are responsible for shaping them?

Bird vocalizations are divided into calls and songs. Calls are short notes, with simple structure, such as being on one pitch. Songs are of longer duration with complex structure, consisting of various notes and pitches. Calls are used to alert young, partners, or group members of potential danger (alert-call) or to keep in touch (contact-call). Both sexes call. The song is given to attract a mate or to announce occupancy or ownership of a territory. In most avian species, only the male sings. For example, if you were to remove a male cardinal from its territory, another male would enter that territory, generally within a day. The male cardinal song is for the other males in the area to hear, as if to say, "This is my spot, and I'm willing to defend it against any and all newcomers!" It is a clear, unambiguous message. California Towhees are monogamous, and once the male attracts a mate and begins nesting, he quits singing. House Wrens are polygynous (they mate with more than one female), and once they acquire a mate and she begins incubating, he resumes singing. Remove a female from a male's territory, and the male will sing more often. These data suggest that male birds also sing to attract a mate. In some species (buntings, thrashers), older birds will have more complex songs than younger birds, and this often assists them in attracting a mate. Most birds use acoustic signals to communicate their species identity, individual identity, motivational state, and physical condition. Because vocalizations are influenced by the environment, to better understand avian communication signals, we have to consider the physics of sound production.

The Physics of Vocalizations

•

Compressed air waves, emanating from a bird singing, spread out much like ripples on a pond. They continue spreading out until eventually they get degraded. In some instances, the sound waves enter the receiver's ears and get compressed. From here, the waves vibrate against the bones of the middle ear, which transmit the vibrations to the inner ear. Fluid in the inner ear is set in motion by the vibrations. Nerves line the inner ear almost like hairs, and these nerves respond to the motion of the fluid (much like grass swaying in the wind) and carry the information to the brain. Depending on which nerves are stimulated in the inner ear, and to what extent, the brain interprets the stimulation as loud or soft, high or low pitched. Birds generally produce sounds in the range of 100 to 5,000 hertz, but there are many exceptions. A hertz is a unit of frequency, where 1 hertz = 1 cycle of a sound wave (crest-crest or trough-trough). A Ruffed Grouse beating its wings rapidly against its breast produces a sound around 80 to 100 hertz, while Blackpoll Warblers and Golden-crowned Kinglets produce sounds between 12,000 and 14,000 hertz. Although we might assume that birds hear about as well as we do, or better, since they vocalize so frequently, recent research suggests many birds may not have as broad of range of hearing as humans.

Imagine two people needing to communicate with each other, but they are on opposite sides of a wide, fast-flowing river. The river is loud, so naturally they shout to be heard. When that doesn't work, they cup their hands over their mouth to direct the sound waves and try again. Again, the river is too loud, and they can't hear each other. Their voices are attenuated, and the message is degraded and never received. The take-home message? Bird calls and songs have evolved to maximize their potential to transmit sound, and they differ in pitch, amplitude, and cadence because birds occupy different habitats that present different acoustic challenges. To a bird living in a dense forest, the trees act as barriers and can degrade air waves quickly, but a bird living in a grassland is faced with a different problem—there are few, if any, tall perch sites from which to broadcast their call. Owls make repetitive vocalizations of short duration and low frequency because those sounds carry farthest in dense forests. Low-pitch sounds with long amplitudes, like the hoot of an owl, bend around large objects (e.g., trees) and carry a greater distance in a densely wooded forest than high-pitch sounds. The amplitude of a low E (bass note, 440 hertz) from crest to crest (or trough to trough) is 30 feet! That is why bass notes from a neighbor's house or someone's car penetrate through walls and car windows. High-pitch sounds have short amplitudes and are bounced around by dense objects; they do not carry far in dense woods. Meadowlarks and Bobolinks make high-pitched (often buzzy) songs of long duration because that type of song travels well in open spaces like prairies and meadows. These high-pitched songs or calls carry in the wind, much like bubbles floating in the air.

In 1980, researchers Sheri Gish and Eugene Morton from the University of Maryland recorded songs from Carolina Wrens in two different habitats to investigate whether the song differed in each. One habitat was open palm/palmetto in Florida, and the other a dense, deciduous forest in Maryland. The researchers played the song from the Florida wren in the Maryland habitat and then rerecorded it approximately 160 feet (about 50 meters) into the habitat. They then played the song from the Maryland wren in the Florida palmetto, again rerecording it 160 feet into the habitat. They found that the songs from the two populations of wrens were both different and differently attenuated. Songs from the Florida wrens traveled farther in the palm/palmetto habitat than in the deciduous forest (the same was true of the Maryland wren's song in the deciduous forest). We can conclude, then, that where a bird lives and the habitat it occupies will largely determine the type of song it will broadcast. Natural selection is responsible for shaping these vocal patterns. When male birds are competing with each other to announce a territory and attract a mate, the one that can produce a song that carries farther in a specific habitat type better than its competitor's will be the one most likely to attract a mate and breed. Over time, the species' song has been refined to reflect the habitat they occupy.

Loon Vocalizations

•

Loons do not sing, at least not in the traditional sense; they vocalize. Technically, birds that sing have a special bone (a pessulus) in their voice box (syrinx) that gives them greater control. Birds that have this bone are called oscines, or songbirds. Loons lack a pessulus. Common Loons produce many different vocalizations, including three in particular that can travel a long distance (more than 600 feet): wails, tremolos, and yodels.

Wails come in three durations: one, two, or three notes. One-note wails consist of a single unbroken note that gradually rises in frequency in the middle and then gradually returns to the original note. Two-note wails begin like a one-note wail, but the first note is shorter, and it quickly moves to a higher frequency on the second note. The second call note may return to the original note or remain on a high note. The average call of the note is two seconds, but this varies from shorter (about one second) to longer (about three seconds). Three-note wails are similar to the two-note wail and have a third note still higher in frequency than the second note (although it can drop and return to the same pitch as the second). The third note is nearly an octave higher than the first note. The duration is typically around three seconds, but again there is variation. Loon wails mean "Come here" or "Here I come." They are also an excellent example of a graded signal. Humans use graded signals all the time. Imagine your four-year-old daughter is backing up at the top of a staircase, getting closer and closer to the edge. At first you might say, "Careful," and if that didn't work, you would raise your voice, and say in a louder voice, "CAREful!" If she continued to where the risk of her falling was imminent, you would shout loudly, "CAREFUL!"

The graded signals of loons improve communication between pair members. Each increase in the signal raises the level of intent. For example, imagine a pair of loons in a territory with two chicks and the female watching over them. The male is feeding on the other side of the territory when some campers walk down to the lake. The female is nervous but not overly alarmed, since the campers on the shore do not pose a great risk to her young. She calls out a one-note wail. The call is short, its message clear, and it travels across the lake because it has a broad frequency. The male receives the call. He stops foraging, looks back at his mate, and is alert but does not leave his foraging area. Fifteen minutes elapse, and the campers get into their boat and make their way toward the loon family. The female is more nervous than before because the boat is a perceived threat. She responds accordingly by producing a two-note wail. Her partner doesn't respond initially because he is underwater foraging, but as soon as he surfaces, he hears the call and responds immediately with a two-note wail of his own. This time, he swims toward her and reduces the distance between them. The boat changes direction and is no longer heading toward the loon family. The female quits calling, and the male returns to foraging. Fifteen minutes later, a Bald Eagle flies over the lake. Worried about her chicks, the alarmed female lets out a three-note wail. The male hears it, answers with a three-note wail of his own, and swims toward his mate. This sequence of wails is an example of a graded signal, in which the sender can alter the message based on the perception of the threat.

Tremolos, another distinct loon vocalization, are often termed "laughing calls." Like wails, they are also graded and can be easily identified as Type I, Type II, and Type III. Each type builds on the preceding type, increasing in frequency and conveying increased motivation and intensity. Tremolos are given when a loon is alarmed, greatly alarmed, and downright about to lose it. Their frequency is modulated; that is, the pitch rises and falls. A modulation is one rise and one fall in pitch. A loon tremolo has between 1.5 to 7 modulations. What is the value, if any, of having a modulated call? To answer that, we can look at other species that have modulated calls, both Red-tailed and Broad-winged Hawks, for example.

Both loons and hawks inhabit areas where air densities can vary substantially. These habitats have warm air currents, called thermals, and below them the air is significantly cooler. A modulated call penetrates pockets of air with different densities and travels a longer distance better than a simple high- or low-frequency call. Vary the pitch, modulate it, and it will travel a greater distance. Loons live on lakes, and when cool air settles in over them, warmer air resides above it. This situation is especially common on northern lakes, where cooler air condenses over the warmer water, forming a thick cloud of mist. I have been lost more than once in such mist trying to catch loons! If a loon wants to communicate its intentions across a lake, through air of varying density, a modulated call will be more effective than an unmodulated call. But there is another twist in the above scenario: the warm air on top of the cooler air creates a channel, where sound waves can be propagated farther (sort of how whale songs can travel across thousands of miles of ocean). Like wails, tremolos have been molded by natural selection and can travel a great distance across a lake.

Some tremolos are produced in a different context. For example, loons may produce a tremolo while in flight. These flight tremolos are most often given by loons flying over another loon's territory, and they may or may not illicit a response from the territorial birds. Pair members will also alternate tremolos calls, most often when a situation makes them nervous, such as at night when other loons are calling, or when an underwater predator attacks a chick. A loon pair duetting indicates a certain amount of angst on the lake. A few birds will duet as part of courtship, but loons' duetting is usually a response to something that has agitated them.

The third call is the yodel. Only male loons yodel. This call is characterized by two distinct parts. The first is a long introductory phrase of three or four notes, increasing in intensity and frequency. The second part is a series of repeated two-syllable phrases. The first part lasts under one second, while the second part can last for several seconds, with each repeated phrase lasting roughly a half-second long. Individual loon yodels are unique, and they can be distinguished by examining sonograms. A sonogram is a graph that plots duration (in seconds) versus frequency (in hertz). By examining a sonogram, we can see graphically how each yodel differs slightly from others. If your hearing is refined and you know what you are listening for, you can tell individual loons apart. For example, Jay Mager, an expert on loon vocalizations, can listen to a recording of a loon yodel and tell you what lake it came from. How cool is that! When a male loon yodels, it extends its neck, crouches low, and swivels in all directions, broadcasting its call. The loon may also yodel from a vulture position, in which it is rises up vertically out of the water, sustained by actively paddling its legs. This überaggressive display occurs only when an intruding loon gets within about 50 feet of a territorial loon. The loon holds its neck and bill horizontally, directly at the intruder. The yodel can travel miles across a lake (close to 10 miles under optimal conditions) and through the woods to an adjacent lake. Who is it directed at, what is its function, and at what time of the day does it occur?

Whereas some male birds sing to attract a mate, the loon yodel appears to be a signature mark of territorial occupation and a readiness to defend it from intruders. Its message and intent are clear; there is no ambiguity: I am a male loon. This is my territory, and I am prepared to defend it! If loons were using yodels to attract a mate, we would postulate that their yodels would occur more frequently before they acquire a mate, and would then drop off during incubation. Yet yodeling does not diminish during nesting, suggesting that it primarily communicates territorial ownership. In May 1993, when Charlie Walcott was with me at Seney, he became the first acoustics researcher to record loon yodels of color-banded loons. We drove around the refuge for days, stopping at lakes with nesting loons. Charlie brought out his sound equipment, including a large parabolic dish, and played a tape of a loon yodel. In most cases, it would illicit a yodel from a male loon, and he would record it. Over a couple of years, he confirmed what had been suspected: that loons on the same territory produce exactly the same yodel for consecutive years. His work showed that male yodels could be discriminated from each other based on a subtle change in the duration of the second and third introductory notes. Charlie also discovered something unexpected: a loon can and does change its yodel if it changes its territory. Previous research (based on unbanded birds) had led researchers to think that the male yodel was stable from year to year. But Charlie's research found that in every case the loon occupying a new territory changed its tune to sound less like that of the previous territorial male. This suggests that the new loon on the lake knew the yodel of the previous territorial holder. Charlie and his collaborators also found that one male changed its yodel when it switched to a different lake but changed back to its original tune after returning to its original lake. This work was the beginning of an exciting stretch of acoustic research in loon behavior.

Jay Mager was Charlie's first graduate student at Cornell to study the many intriguing yet subtle aspects of loon vocalization. Jay worked tirelessly from his canoe, recording yodels and investigating regional differences among hundreds of banded loons across North America. He further studied the two parts of the yodel that travel farthest: the peak frequency and duration of the yodel, which includes the third note of the introductory phrase; and the number of repeat phrases. These features likely carry information, but what kind? To find out, Jay played back recordings of yodels from different loons (neighbors on the same or adjacent lake; strangers from an area at least 20 miles away) and observed the behavior of the loon hearing the recordings. The work required patience and long hours, and he was sometimes (not so gently) reminded by his subjects that he was in their territory. Once he was in his canoe doing a playback experiment when a highly agitated loon attacked—it repeatedly struck and stabbed at the canoe, to the point of damaging it!

The following is just some of what Jay discovered. First, larger males tend to produce lower-frequency yodels. For example, male loons from Minnesota have higher-pitched yodels than loons found in New England, which tend to be larger. Jay found that he could distinguish which region of the country a loon lived in from hearing its yodel. Second, a smaller loon that emits a high-pitch yodel upon hearing a playback of a low-pitched yodel from a larger loon will respond by changing its yodel. Since lowering pitch involves some physical constraints, the loon increases the number of repeats in the second part of the yodel. Weight and condition are related to the ability of a male to acquire or defend a territory; thus, the lower-frequency yodels produced by larger loons may signal a greater threat to neighboring loons. By increasing the duration of the second part of the yodel, the loon is sending a message conveying intent and motivation: I am ready to fight and defend my territory. Jay found that territorial pairs respond with more aggression to playback yodels of strangers having more repeat syllables. Furthermore, males can discriminate between yodels from territory neighbors and nonneighbors, and females can likely discriminate yodels from their mates and nonmates.

The best, or clearest, communication occurs when there are fewer background noises, and at a time of day when attenuation of the signal is weakest. Selection operates to maximize communication between species members because it saves everyone time and energy, important elements in the daily grind of survival. Remember, wind is an important factor in the degradation of a signal; too much of it presents acoustical challenges. Given the warming heat of the sun, winds pick up during the day and settle down at night. Although we might predict, then, that loons yodel or call more during the night than during the day—that yodeling is more a nocturnal than diurnal behavior—many campers and residents of lakeside cabins can attest to the fact that loons vocalize day and night. But their calls are due to different factors during the day than at night. During the day, loons give long-distance communication calls (wails, tremolos, yodels) primarily in response to human-related disturbances (e.g., canoeists, kayakers, boats, jet skis) and aerial threats (e.g., Bald Eagles, low-flying aircraft). At night, many of these factors are removed from their territory, and loons vocalize to communicate primarily with other loons in the neighborhood. Intraspecific communication (neighbor–neighbor, neighbor–stranger) occurs predominately between 11:00 p.m. and 1:00 a.m., which indicates that a loon's assessment of its landscape (Who are my neighbors? Are there any strangers?) takes place more at night than during the day.

Who would have thought that the yodel, a single vocalization, could provide information necessary for other loons to assess the identity, competitive ability, and aggressive motivation of the signaler? We have come a long way in understanding the significance of long-distance vocalizations, but there are more questions to be answered. Do sons and fathers have similar yodels? How much of the yodel can be attributed to genetics, and how much is social? Do loons' yodels in certain neighborhoods have components distinct from those in other neighborhoods? What feature(s) are they using for individual recognition? Can they recognize kin, and do they respond to them differently than to unrelated individuals? Science has reached a new era where the answers to these questions are within reach. It is indeed an exciting time to be an acoustic loon researcher.

Short-Distance Communication

Loons make a number of calls in the short-distance communication category, such as hoots, mews, and toots (loon chicks make additional begging calls). Because the distance between the sender and receiver of these calls is minimal, attenuation is not a factor in shaping them, but selection is still at work. Perhaps the following phrase sums up short-distance communication calls in loons: "I'm standing right next to you, you don't have to shout!" Selection appears to have favored the conservation of energy: why make a loud call when a soft one will do? Hoots are short, single notes, as the name suggests, given by individuals that are near to each other. They essentially serve as a contact call, a checking in with their partners during periods of low aggression or anxiety. Therefore, they are most often given and heard between mates but can also be heard when a loon approaches a group of other loons, as they often do in late summer and during fall. Mews are mostly restricted to between mated pairs during courtship and are often used by one adult to lure its mate to land for copulation. Parents also mew to communicate with their young that it is okay to come out of hiding. Toots are similar to hoots, but the beginning and ending sounds are harsher than a hoot (think hard versus soft consonant). They are most often heard between mated pairs, usually after a disturbance.

Do All Loons Sound Alike?

Yes and no. Yes, in that wails and yodels are produced by all five species of loons, and no, in that only Common and Yellow-billed Loons produce a tremolo and only Red-throated, Pacific, and Arctic Loons produce yelps, a type of short-distance communication call. Red-throated Loons have the most atypical vocalizations of all loon species. They are the only species to produce a croak call (though they do not hoot) and the only one in which both sexes produce a yodel. Given this, if behavior (in this case, a vocalization) can be used to infer evolutionary relationships, then it appears that Red-throated Loons split first from the other four species. The other four loons are united because they all produce hoots. Common and Yellow-billed Loons are united and split from Pacific and Arctic Loons because they alone produce a tremolo. Similarly, Pacific and Arctic Loons are nearly indistinguishable from one another, just as Common and Yellow-billed Loons are. The diversity and similarity of vocalizations produced by loons support both the molecular and fossil data, which suggest that Red-throated Loons are the oldest of all the loon species, and the next evolutionary split separated Pacific and Arctic Loons from the Common and Yellow-billed Loons. Biologist Alec Lindsay's 2002 dissertation at the University of Michigan produced much of this information. He further suggested that the ability of Common and Yellow-billed Loons to produce tremolos came from a genetic change that modified the syrinx, allowing their production as an extension of wails.

If the ability to yodel was ancestral, why did the females of loon species other than the Red-throated lose this behavior? First, think about the purpose of the yodel in female Red-throated Loons. Remember, they nest on small ponds or lakes, much smaller than those that other species of loons use. Because they cannot obtain enough food on their nesting pond to support their developing and growing family, they leave the pond often to forage in the ocean, leaving one member of the sex behind with the young. Alone with the young half the time, the female yodels to minimize the probability of other loons landing nearby. The function of the female yodel, then, is the same as the male yodel, territorial defense. The reason that descendant female loons dropped that vocalization could be that females of the other loon species are not as involved in defense of the territory and thus the yodel was unnecessary for them. But Walter Piper (among others) found that Common Loon females were just as active in territorial defense as males. In fact, females of the other four loon species are involved in territorial defense because the lakes they utilize are so much larger than the small ponds and lakes that Red-throated Loons use. However, since males never leave the lake during the pre-nesting and nesting periods, females can defer the yodeling to them.

Other aquatic fish-eating birds, such as grebes and cormorants, do not make as impressionable vocalizations as loons. Why the difference? Birds in the same order, family, and genus have similar vocalizations due to shared similarity of the size, thickness, and location of the syrinx and associated musculature. Thus, all loon species sound more like each other but different from grebes. There is likely some convergence of songs based on the occupation of similar habitat types, but there seem to be genetic (evolutionary) constraints. For example, I will never sing soprano, but I do sound like my brothers on the telephone. Large cats, including lions, tigers, leopards, and jaguars, roar, and because they share this feature, they are grouped collectively in the genus *Panthera*. Cougars (mountain lions) are large cats, too, but since they cannot roar, they are not placed in the *Panthera* genus. The reason they cannot roar is that they lack a ligament in the voice box that stretches to produce a wider range of sounds. The cougar's inability to roar is an example of a genetic constraint. All five species of loons produce similar vocalizations and as such are lumped together in the same group, while grebes and mergansers have different calls (and no male song) and are therefore placed in their own respective groupings. I suspect that the life history characteristics of loons that we know so well developed in large part because of their ability to produce different vocalizations early in their evolutionary history.

We enjoy listening to loons call. Is this because their calls are so recognizable and distinctive from other bird vocalizations, or is it because these calls have identifiable structure and rhythm that we can relate to? We love repeated patterns (the Beatles adopted this strategy), and the yodel, tremolo, and wail all have repeated notes and patterns. Or does our enjoyment come from something deeper? Music is primal, an integral part of all human cultures. Perhaps our vertebrate brains were prewired for song detection long before humans evolved. This would, in part, explain why we find so much emotion and meaning in music and in the vocalizations of loons, and why we remember them decades later. Those haunting and familiar loon calls are reminiscent of wild places that invigorate us spiritually and help us gain perspective on what is really important in life. Maybe those calls exist to remind us to slow down, enjoy the moment, value our time with loved ones and friends—to just sit and listen.

Further Reading

Barklow, W. E. 1979. The function of variations in the vocalizations of the Common Loon *(Gavia immer)*. Ph.D. diss., Tufts University.

Bradbury, J. W., and S. L. Vehrencamp. 2011. *Principles of Animal Communication.* 2d ed. Sunderland, Mass.: Sinauer Associates.

Catchpole, C. K., and P. J. B. Slater. 2008. *Bird Song: Biological Themes and Variations.* Cambridge: Cambridge University Press.

Endler, J. A. 1993. Some general comments on the evolution and design of animal communication systems. *Philosophical Transactions of the Royal Society of London B* 340: 215–25.

Gaunt, A. S. 1983. A hypothesis concerning the relationship of syringeal structure to vocal abilities. *The Auk* 100: 853–62.

Gish, S., and E. Morton. 1981. Structural adaptations to local bird habitat acoustics in Carolina Wren song. *Zeitschrift für Tierpsychologie-Journal of Comparative Ethology* 52: 74–84.

Lindsay, A. R. 2002. Molecular and vocal evolution in loons (Aves: Gaviiformes). Ph.D. diss., University of Michigan, Ann Arbor.

Mager, J. N., and C. Walcott. 2007. Structural and contextual characteristics of territorial "yodels" given by male Common Loons *(Gavia immer)* in northern Wisconsin. *Passenger Pigeon* 69: 327–38.

Mager, J. N., and C. Walcott. 2014. Dynamics of an aggressive vocalization in the Common Loon (*Gavia immer*): A review. *Waterbirds* 37: 37–46.

Mager, J. N., C. Walcott, and D. C. Evers. 2007. Macrogeographic variation in the body size and territorial vocalizations in male Common Loons (*Gavia immer*). *Waterbirds* 30: 64–72.

Mager, J. N., C. Walcott, and W. H. Piper. 2007. Male Common Loons, *Gavia immer,* communicate body mass and condition through dominant frequencies of territorial yodels. *Animal Behaviour* 73: 683–90.

Mager, J. N., C. Walcott, and W. H. Piper. 2010. Common Loons can differentiate yodels of neighboring and non-neighboring conspecifics. *Journal of Field Ornithology* 81: 392–401.

Mager, J. N., C. Walcott, and W.H. Piper. 2012. Male Common Loons signal greater aggressive motivation by lengthening territorial yodels. *Wilson Journal of Ornithology* 124: 74–81.

McIntyre, J. W. 1988. *The Common Loon: Spirit of Northern Lakes.* Minneapolis: University of Minnesota Press.

Mennill, D. J. 2014. Variation in the vocal behavior of Common Loons (*Gavia immer*): Insights from landscape-level recordings. *Waterbirds* 37: 26–36.

Piper, W. H., K. B. Tischler, and M. Klich. 2000. Territory acquisition in loons: The importance of takeover. *Animal Behaviour* 59: 385–94.

Walcott, C., D. C. Evers, M. Froehler, and A. Krakauer. 1999. Individuality in "yodel" calls recorded from a banded population of Common Loons, *Gavia immer. Bioacoustics* 10: 101–14.

Walcott, C., J. N. Mager, and W. H. Piper. 2006. Changing territories, changing tunes: Male loons, *Gavia immer,* change their vocalizations when they change territories. *Animal Behaviour* 71: 673–83.

Wentz, L. E. 1990. Aspects of the nocturnal vocal behavior of the Common Loon (Aves: *Gavia immer*). Ph.D. diss., Ohio State University.

Young, K. E. 1983. Seasonal and temporal patterning of Common Loon, *Gavia immer,* vocalizations. Master's thesis, Syracuse University.

7

A Vigilant Bird

Parents, Chicks, and the Breeding Loon Family

Around ten o'clock on the evening of June 30, 1993, I was leading a discussion with my Earthwatch volunteers about a loon pair we had observed for the past sixteen hours at Seney National Wildlife Refuge. We had watched them from a half hour before sunrise to a half hour after sunset, and even after a long day in the field there was excitement in the room. The first group noticed that the female loon did not want to leave the nest when the male came to replace her. She appeared fully committed to staying there, and only when the male climbed up the nest bowl and nudged her did she decide to leave. The second group noticed a similar scene, but in this case the roles were reversed: the male did not want to leave. These were telltale signs that something was up. The last group of the day observed that even though the female was on the nest incubating, the male was active nearby, vocalizing frequently with territorial yodels. Why was it doing that? they asked me. I offered the suggestion that perhaps the male could hear the chick calling from inside the egg and, like us, was excited for the egg to hatch. Based on the behavior of both adults over the past twenty-four hours, I speculated the egg(s) would likely hatch the next day or the one following. Personally, I was anxious because my research was centered on documenting and understanding parental roles during the chick-rearing phase. Specifically, how do the sexes divide the workload? Does one sex spend more time watching, brooding (protecting them under their wing), or allowing the chicks to ride on their back more than the other? Does one sex feed the chicks more than the other?

The following morning, at 10:03 a.m., the first chick hatched, as I predicted, though I was not there to see it for myself. Instead, the second shift, led by my assistant, Ari, documented its arrival. The chick remained on the nest for twenty minutes before plopping down into the water. It stayed close to one of the parents throughout the day, sometimes under the protection of the adult's wing. It swam about and later in the day even climbed back into the nest. That evening, the male yodeled frequently, as had occurred the previous night. I prepared my volunteers during our daily discussion for the possibility of moving to a new viewing area once the second chick hatched. Loons often move their chicks to a different spot on the lake from where they were born. These areas are called nurseries because the family spends most of its time there. Nurseries are typically protected bays or coves, out of the prevailing winds, with an ample supply of minnows and young fish.

We went over logistics, then I posted the schedule for the following day. Ari and I switched shifts. The next morning, he left with the first team at 4:30 a.m. By 6:00 a.m., the second chick was visible and within thirty minutes of hatching made its way to the water's edge. When my shift showed up around 9:00 a.m., the chicks were on the back of the male, and the female was off foraging. The female returned at 10:45 a.m., and the chicks slid off the male and swam between the two adults. About an hour later, the older chick pecked the head of its sibling for the first time. It was hardly much of a blow, but we recorded it and kept observing. At 3:12 p.m., Ari's group observed the older chick peck at the head of its sibling several more times. The topic of sibling rivalry came up during our discussion later that night, and though it is uncommon in loons, there was no mistaking that is what we had observed.

The same behavior happened the following day, except the pecking occurred frequently throughout the day. An Earthwatch volunteer asked me, "Is there anything we can do?" This was a teachable moment. Nature can be cruel, heartless, and indifferent; its rules are not governed by fairness. Eat or be eaten. This was a science investigation, and we were there to observe and document. I thought it best not to interfere, despite the emotional trauma of observing a young chick being harassed by its older sibling. Maybe it would have a happy ending. Yet day after day, the older chick grew in strength, and the blows to the younger sibling came with more force. The younger chick fell behind its parents and older sibling as they moved around the lake. It was constantly calling out, producing begging calls, but neither parent seemed to notice; they appeared indifferent. On a couple of occasions, one of the parents swam back and attempted to feed the chick, but the intervals between feeding periods lengthened. Worse yet, between feedings, the older sibling would swim back to the younger one and peck at it some more. On day eight the young chick was no longer able to keep up with the family moving around the lake, and the parents had stopped making any effort to feed it. On day nine, late in the morning, the older chick swam back to its younger sibling and struck its skull with forceful blows from its bill. At this point, the young chick was no longer a competitor for food, so we wondered why its older sibling was keeping up its relentless domination. Every day for the next ten days the blows to the skull continued, and on day eighteen the older sibling pecked continuously at the young chick's head for fifteen consecutive minutes. Finally, the younger sibling sank and never resurfaced. It died at 2:43 p.m.

Sibling rivalry is normal among some species of birds, such as egrets and eagles. In each case, the young hatch asynchronously, creating a situation where one sibling is older and more dominant, often bullying its younger sibling. In some cases, the older sibling kills its sibling as a form of brood reduction, ensuring it alone gets the food provided by the parents. Siblicide, or cainism, is uncommon in loons. The dominance hierarchy between loon chicks is established by the second or third day and continues throughout the fledgling phase. The older chick will always weigh more than its younger sibling, but in the great majority of loon pairs there is no overt aggression, and, unless a predator takes one of the chicks, both chicks will fledge successfully. What triggers siblicide in loons is unclear, but based on my few observations and experiences, it appears to be genetic rather than environmentally induced. Since the lakes where I observed siblicide had plenty of minnows, shortage of food could not have been the reason. Though unusual, siblicide is one example of the complex range of behaviors related to a loon family unit.

Chicks at Birth

•

Upon hatching, loon chicks are fully feathered but far from independent. They cannot feed themselves, and without their parents' assistance, they would starve in a matter of days. Being surrounded by water continuously, chicks face thermal challenges and need support in the form of parental brooding, such as a shielding wing or a hop on the back of one of the adults. Natural selection operates on stages of chick development in birds. Some chicks are born naked and helpless, a condition known as altricial, while others are born fully feathered and capable of feeding themselves, precocial (or superprecocial). Loon chicks fall somewhere near or just above the middle of this range (a condition known as subprecocial). Survival would be impossible for loon chicks if they were born altricial, and based on what we know of development in the egg, it would be near impossible to hatch a chick with the musculature and nervous systems developed to the point where it could catch fish independent of its parents. Subprecocial hatching works just right for loon chicks. Loon parents have to be attentive to their young, especially during the first two weeks of their life when they are still developing muscle and motor coordination to swim and dive. Parents also have to protect their young from aerial and underwater predators and help them maintain their core body temperature.

Most loon eggs hatch asynchronously. The second egg generally hatches eighteen to twenty-four hours after the first, though there are always exceptions. This asynchrony occurs because it takes the female roughly fourteen to twenty-four hours to pass the second egg, so the chick inside the first egg has a slight developmental head start. If the second chick is born much later than the first one, it will be that much further behind. Some studies have shown that the first sibling born from a two-egg clutch is larger and stronger and has the upper hand compared to its younger sibling; it also has a higher survival rate. The second egg is insurance should one of the eggs be sterile. This pattern of egg laying is observed in many species of birds, not just loons. Laying a single egg is literally putting all your egg(s) in one basket and is certainly not a safe evolutionary strategy. A one-egg clutch is rare in the bird world, observed in only one group of birds, the tubenosed seabirds such as albatrosses and petrels. It is better to lay a second egg, even if abandoned later, since the time invested in mating has already taken place.

Plumage Changes in Loon Chicks

•

Young chicks are born with black downy feathers. An adaptationist would surmise that black confers an advantage over another color because of its absorptive properties. The belly is countershaded (see chapter 2) but not to the extent observed in adults. At around ten to fourteen days, the black feathers are pushed out and replaced by brownish-gray down feathers. In aquatic species like loons and grebes, the first thick and long feathers to develop occur on the ventral side, which helps insulate chicks from cold water. In many upland species, such as grouse and Killdeer, the first developing feathers grow on the back because these birds are more vulnerable to losing radiative heat upward. Why brownish gray and not black? I suspect more melanin is needed to produce the black color than brown, and since the production of melanin takes nutrients and energy, less energy would be available for growing, which is what the chick is doing at this time. By days ten to fourteen, the young chicks have grown substantially and are better able to maintain and regulate their body temperature. In another two weeks, around days twenty-six to thirty, the puffy brown down is replaced and pushed out by gray body (contour) feathers. Over the next month, the body and flight feathers replace all the down. Thus, loon chicks undergo two molts prior to fledging, which is rare in the avian world (only penguins share this trait).

Chick Development

•

I have observed loon chicks enter the water within one hour of hatching. By the second day, they are peering into the water. During the first week, they are pursuing insects skittering on the water's surface, and making their first dives. Beginning around day seven, they dive more frequently and begin to look for underwater food. And by the end of week two, chicks rapidly become more adept at swimming and chasing fish underwater, though more often than not, they are unsuccessful in capturing them. In 1970 and 1971, biologist Jack Barr filmed, swam, and studied loon chicks in pens situated in Ontario's Algonquin Provincial Park. He observed that chicks at three weeks of age can pursue fish underwater for up to 100 feet but were still rarely successful, only 3 percent of the time. Adults feeding chicks of this age, or older, typically dropped fish in front of them, allowing them to pursue and capture the injured prey. By week six, the capture success rate for a chick on its own was still only 14 percent. Normally, once chicks reach six weeks of age, they have a 90 to 95 percent chance of fledging and flying off the lake in the fall. By week nine, their capture success rate was 51 percent, but parents still provided them with fish. By week eleven or twelve, they were mostly independent of their parents, often swimming throughout the day by themselves.

The growth of loon chicks exhibits the typical S-shaped pattern: it is slow for the first few days, rapidly accelerates and increases exponentially for the middle part of their development, and then as they near fledging, it slows and plateaus. Kevin Kenow, a biologist with the Upper Midwest Environmental Sciences Center (USGS) in La Crosse, Wisconsin, calculated that the maximal growth spurt in loon chicks occurred at day twenty-four (between weeks three and four), and by day sixty-six the chicks had reached their near maximal body mass (about 3,250 grams, for a chick in Wisconsin). Working with François Fournier, Kenow found that parts of the body grew at different times. For example, the lower leg bone, the tarsus, grew more rapidly than many other structures (e.g., bill, wing) and reached its plateau around day forty. Because the leg is so important in swimming and critical to a loon's survival, it is no surprise that it develops sooner than other body structures. By week eight, a chick's webbing and toes are nearly the size of the adult's, and the bill is about 75 percent of its adult length. The bill does not reach near adult size until week eleven or twelve (a longer bill likely increases foraging success). Fournier and Kenow measured daily energy expenditure at days ten, twenty-one, and thirty-five. Their results were a bit surprising. They found that ten-day-old chicks used as much energy as twenty-one-day-old chicks. This result was atypical because normally the bigger the individual, the more calories it burns and needs. They surmised that since loon chicks spend nearly all their time in water, which robs them of heat, they need to eat more than would be expected for their body size in order to compensate for the heat loss.

Parental Care

•

What is best for the chick is not always in the best interest of the parent, and what is in the best interest of the parent is not always best for the chick. Parents brood their young, shelter them from heat or cold, and defend them against would-be predators. They provide them with food, which for birds can be very demanding. A House Wren makes 491 feeding trips in one day. To feed its young, a Chimney Swift flies 600 miles in one day. Barn Owls will bring a rodent to their young ten times during the night. A hungry loon chick will keep begging and calling for food, and the parent will keep providing it, but there is a tipping point at which the parents' self-interest and self-preservation override that of their chicks. The bottom line is there must be enough food and time in the day to procure it, not only for the chicks but for the adults as well. Behavioral ecologists have long noted this existing tension between parents and offspring (known as parent-offspring conflict). Fortunately for loons (and most birds), both parents share in the responsibilities of caring and providing for their young. But this brings up another question: how much care should each pair member provide?

We can argue that the parent providing more parental care may be more disadvantaged than the one providing less care because the former will have less time to forage for itself than the latter. Depending on how much less time, that may be the difference between meeting or not meeting its daily caloric needs. Because of this, tension exists between pair members that provide care to their young. Is everyone pulling an equal weight? The more care the chicks require, the greater the tension between the parents, and the more they have to cooperate and communicate if they are going to successfully raise their young. Because loons are socially monogamous and form long-term pair bonds, we could expect that cooperation would evolve and that behaviors would be finely tuned to each other. Pair members should cooperate and compensate for each other to ensure both current and potential future reproductive success. Flexible parental roles allow each pair to adjust their amount of care as ecological, environmental, or social conditions change, and to spend more or less time using specific behaviors that are best suited to each pair. If this is the case, then we would expect parental care to vary to some degree among pairs.

From 1993 until 1996, my Earthwatch teams and I observed eleven loon pairs with their chicks for 2,414 hours. Because each parent was banded, I could distinguish males from females, but not all the time. In many cases I could not see the bands and could not distinguish between the parents. To solve this problem, I purchased some model paint and put a white vertical stripe on the loon's lower bill. This allowed me to tell from a considerable distance which pair member was responsible for doing what. I defined behaviors associated with parental effort as time spent feeding chicks, resting with young (vigilance), and defense of young or territory (agonistic), and I summed those up for each pair member and compared them across all of the loon pairs. I found no significant difference in behaviors between males and females, though males spent slightly more time resting with chicks than females. This was more common during the first two weeks. There was a high degree of intrapair and intrasexual variation. If parental effort was shared equally (fifty-fifty), then time allocated for those behaviors should be comparable among sexes and pairs. If it deviated from fifty-fifty, then we could reasonably expect to observe differences in parental effort among sexes and pairs. What did I observe? In six pairs the females made more effort, in three pairs the males made more effort, and in two pairs parental effort was approximately fifty-fifty.

The observation that parental effort varies among pairs and that no one sex has a more dominant role in parental effort than the other suggests that each pair develops its own coordination of behaviors to maximize reproductive success. Parental roles appear to be dynamic in loons, and this flexibility may be beneficial in species that obtain new partners over the course of a lifetime. Because sex roles are not rigid, each individual can adjust the care given relative to its mate's, and these adjustments may maximize the pair's current reproductive success. Loons should choose mates that cooperate in territorial defense and have good parenting skills, such as a willingness to incubate and pay attention to their chicks. Everyone's role and contribution to the partnership may depend on territorial quality, predator abundance, age, experience of individuals, and, finally, the success of the partnership itself.

Providing Food for the Young

In nine of the eleven pairs I observed, females delivered more prey to their chicks than their partners, while males spent more time resting with young than females (vigilance), although there was considerable sex and pair variation. Females delivered more prey than males but did not spend more time feeding the chicks. This suggests that females are more efficient at pursuing prey and perhaps feeding different prey items than males. Because the male loon's bill is longer, deeper, and wider than the female's, smaller prey items, such as minnows, may be more challenging for male loons to acquire. I observed no difference in the duration of successful underwater foraging dives between males (33.6 seconds, n = 3,806) and females (33.1 seconds, n = 4,381).

Parental effort showed a significant decline for both sexes during the chick-rearing stage. As the chicks matured and gained in their ability to dive, swim, and forage independently, the parents fed them less often. On occasion they brought large (sometimes too large) fish back for the chicks to swallow. At times the situation was a bit comical since some fish were nearly the size of the chick! When the chicks required less provisioning and periods of resting, both parents could spend more time foraging for themselves and in some cases take short flights.

Pairs that coordinate parental duties during chick rearing (i.e., alternating between vigilance, chick feeding, and self-foraging) are likely to raise more young than pairs whose members act out of self-interest. Four of the original eleven pairs (36.4 percent) that I studied lost their broods within fifteen days after hatching. In addition, one of the three replacement pairs also lost their young. The causes of these mortalities remain unknown, except in one case in which I observed a Bald Eagle fly over the loon territory and grab one of the chicks that was hidden close to shore. Researchers have long suspected that in addition to eagles, large fish, such as a northern pike, or even a snapping turtle may take a loon chick. Such a high level of chick mortality suggests that coordination of vigilance by both parents is needed to successfully fledge young. Although less energetically costly than chick provisioning, vigilance is a key component of reproductive success, and its importance should not be underestimated when evaluating parental effort. In my observations, both sexes traded vigilance bouts during chick feeding, but there was considerable variation among pairs. Vigilant adults wailed in response to the sight of a Bald Eagle, which supports the idea that increased vigilance improves the long-distance detection of predators.

Food Resource Partitioning

Throughout the avian world, if males and females are the same size and have the same bill measurements and wing length, then they may be direct competitors for the same food, especially if they are restricted to a well-defined territory. Division of food resources among other birds is well established. Several studies have shown that habitat and prey partitioning broadens the food base, reduces competition between pair members, reduces pressure on individual food patches (concentrations), and allows for greater recovery time of the patch. Fish in a lake are likely to be patchily distributed, and their availability likely varies by day and season. To reduce competition and increase their foraging success, loon parents may selectively prefer different sizes or types of prey items. Jack Barr has found that stomach morphology and histology (tissue type) differ between male and female loons. In addition, he found differences between prey type and abundance in the stomach of thirty loons. Collectively, the data suggest there may be enough differences between male and female loons to support the notion of food partitioning.

The Loon's Diet

•

Loons eat fish. But do they have a preference? Experiments on captive loons by Barr showed that if given the choice, a loon prefers to accept species with small heads, long slender bodies with soft scales, and no spines, such as whitefish, cisco, and brook trout. While loons may have a preference for certain fish, that preference may not be available in abundance and under the optimal conditions for capture. Furthermore, just because a species is present does not mean a loon will eat it or selectively go after it. To avoid predation, trout swim with a burst of speed in a direct line toward deeper, darker water. Perch, on the other hand, attempt to elude would-be predators by darting left, then right, in a zigzag pattern. Because loons maneuver well underwater, they may catch perch more easily than trout. In lakes in the upper Midwest, yellow perch is the most common species found in the stomachs of loons, suggesting it is an important prey item.

Loons are predators, and unless food is abundant, which it rarely is, they have little choice but to be opportunistic and expand their diet beyond fish. In the summer, they will eat crustaceans (such as crayfish), insects (dragonfly larvae), and leeches. A couple of loon enthusiasts have seen them swallow mallard ducklings. The diet of wintering loons has been little studied, though biologist Ken Wright has observed them eating shrimp, crabs, marine worms, and even an octopus! Similarly, they may go after fish from solitary, bottom-dwelling (benthic) species such as bullheads (in summer) and flounder and halibut (in winter), to surface schooling fish such as herring and shad (in summer), to menhaden and alewives (in winter).

Compensation and Cooperation

•

The exhibition of behavioral flexibility (a pair bond member's response to the other) should theoretically compensate for a situation in which one member is injured or ill. The notion of compensation between pair-bond members is of great interest to behavioral ecologists. In 1995 my Earthwatch team and I were able to gather some opportunistic data on the topic. We were doing all-day observations of a pair of loons at Merkle Lake, continuous with the Turtle Flambeau Flowage, in northern Wisconsin. The pair had two chicks, both four weeks old. I had caught the male a few days earlier and marked the lower bill with two white vertical lines. On July 24 the adult pair delivered 184 prey items to their two chicks, 55.2 percent of which were delivered by the male. By chance, this pair was randomly selected to be observed again, so we started at 5:00 a.m. and worked all day on July 25. We saw a noticeable difference. The female was not as engaged in chick provisioning as the male. In fact, 80.6 percent of the 175 prey items were delivered by the male. This compensatory increase occurred because the female spent most of the day self-feeding alone on the other side of the territory. She would return to the family unit throughout the day but never remained long. At the end of the day she was united with the family. We did not observe the pair on July 26 but resumed observations the following day. We recorded 171 prey items delivered, 51.9 percent delivered by the male and 48.1 percent by the female. Everything had returned to normal. We gathered more chick provisioning data on this pair for some days in August, and found that the male and female were always within 6 percent of each other, suggesting what we observed in late July was an outlier, compensatory behavior by one pair bond member in response to a lack of chick provisioning by the other pair member. We could never assess what, if anything, may have been wrong with the female.

Infanticide

•

What do loons, langurs, and lions all have in common? To answer that, we have to go back to 1994, July 23, at 2:04 p.m., when I was observing a family unit with two chicks on Manistique Lake in Michigan's eastern Upper Peninsula. The adults were resting roughly 20 to 30 yards away from the chicks when I observed a separate adult loon rise and peer around like a periscope, with just its bill and head, directly in front of me. It dove immediately and surfaced again along shore but much closer to the loon family. Was it foraging? What was this loon doing? It dove suddenly, and I looked intently for it, scanning the surface for any signs. A few minutes elapsed, and then chaos ensued. The parents rushed back toward the chicks: one was missing. I spied the intruding loon with the chick in its bill. *What just happened?* It was difficult to be sure what happened next, but it looked like the rogue loon released the chick and surfaced without it some distance away from the family unit. One of the pair members pursued it, wing rowing and calling across the lake. I never did see the rogue loon fly off from the lake during my observation period, but the lake was large enough to support several loon pairs. I believe that this rogue loon had attempted to commit an act of infanticide.

Infanticide refers to the killing of the young of the same species. Lions and langurs are the more notable mammals that will occasionally exhibit this behavior. To understand why a lion, langur, or loon would attempt infanticide, we have to rely on evolutionary and ecological theory. If a male lion or langur takes over a group of females with young, they will often kill the offspring sired by the former male. This sexually selected infanticide appears to cause the females in the group to become sexual receptive, that is, available and willing to mate again. This works out well for the new male because by mating with the females, he will increase his fitness by siring new young, who will have his genes, not the previous male's genes. In several species of primates, it is believed that 30 to 40 percent of all infant mortality is due to infanticide. The drive to reproduce is strong. For example, Hanuman (or gray) langurs of South Asia live in bands of one large male and a group of smaller females and their offspring. Males compete for band leadership, and should one male usurp the resident male, he will attempt to kill the infants to induce the females to become sexually receptive, thereby increasing his chances of reproducing and passing on his genes. Naturally, because raising their current young is in the females' best interest, they fight back and do their best to prevent infanticide, but they are not always successful.

In birds, infanticide has been documented in Barn Swallows, House Wrens, and Wattled Jacanas. Jacanas are tropical waterbirds that breed in wetlands. They have an unusual polyandrous breeding system in which a female may pair with up to four males simultaneously. The female lays the egg, but each male will incubate and look after the brood when they hatch. Because the young develop quickly, the male jacana does not have to provide much parental care. Occasionally fights happen between females for territory, and should a takeover occur, it is usually when males are caring for eggs or the chicks of the former female. In the mid-1990s, Stephen Emlen of Cornell University conducted a field experiment to test the sexually selected infanticide hypothesis using the adaptationist approach. His team marked thirty-one adult jacanas in a wetland in Panama and attempted to induce infanticide by experimentally removing two breeding females. Replacement females were present on the vacated territories within one hour of the following dawn. Emlen found that each of the replacement females aggressively attacked the chicks of the former female resident. Naturally, the males defended their young, but female jacanas are larger and dominated the males, leading to presumed deaths of two broods of chicks (observing the actual event leading to the death of chicks is challenging). All incoming females initiated sexual behavior within forty-eight hours of killing or driving off the males' offspring. The adaptationist approach predicts that in a mating system where sex roles are reversed (polyandrous), infanticide will be carried out by females, not males. Emlen showed that this was indeed the case, further strengthening the sexually selected infanticide hypothesis by extending it to a species in which sex roles are reversed.

The rogue loon I observed likely did not have a territory or a mate and was trying to break up the pair bond by killing the chicks. Success would have increased the chance that the pair bond would dissolve. Loons that have lost their clutch, or their young, are more likely to switch partners than loons who have not. So, this strategy of infanticide, as bizarre as it may seem, could be advantageous some of the time for the bird who initiates it. Although infanticide is uncommon in loons, there is more we need to know. For example, can researchers identify the genes responsible for this behavior, presuming, of course, that genetic makeup is at least partly responsible? What social conditions, if any, must exist before a loon will commit infanticide?

Birds behave. They eat, sleep, establish and defend a territory, court a mate, form a partnership, build a nest, sit on their eggs, and when they hatch, provide for their young. Throughout this chapter I have discussed several loon behaviors that likely have a genetic basis. These behaviors as expressed are linked directly to the individual's strands of DNA, located in each of its cells. Nature has selected for behavioral traits in loons that maximize their chance of survival and reproduction. Loons share the same DNA that allowed their great-grandparents to be successful. In species such as loons that form seasonal partnerships in which both parents provide for their young to ensure they fledge successfully, cooperation will be selected over selfish behaviors. Behavioral flexibility in contrast to rigid parental roles also aides loon pairs in successfully raising young. However, selfish behaviors, such as siblicide and infanticide, though not common, can also occur. These selfish behaviors exist because the individual benefits from such action. Loons exhibit a variety of behaviors that keep behavioral ecologists on their toes.

Further Reading

Alcock, J. 2012. *Animal Behavior: An Evolutionary Approach.* 10th ed. Oxford: Oxford University Press.

Alexander, G. R. 1977. Food of vertebrate predators on trout waters in north central Lower Michigan. *Michigan Academician* 10: 191–95.

Barr, J. F. 1973. Feeding biology of the Common Loon (*Gavia immer*) in oligotrophic lakes of the Canadian Shield. Ph.D. diss., University of Guelph, Ontario.

Barr, J. F. 1986. Population dynamics of the Common Loon (*Gavia immer*) associated with mercury-contaminated waters in northwestern Ontario. Occasional Paper 56. Canadian Wildlife Service, Ottawa.

Barr, J. F. 1996. Aspects of Common Loon (*Gavia immer*) feeding biology on its breeding ground. *Hydrobiologia* 321: 119–44.

Bent, A. C. 1919. Life histories of North American diving birds. *U.S. National Museum Bulletin* 107: 47–62.

Clutton-Brock. T. H. 1991. *The Evolution of Parental Care.* Princeton, N.J.: Princeton University Press.

Emlen, S. T., N. J. Demong, and D. J. Emlen. 1989. Experimental induction of infanticide in female Wattled Jacanas. *The Auk* 106: 1–7.

Emlen, S. T., P. H. Wrege, and N. J. Demong. 1995. Making decisions in the family: An evolutionary perspective. *American Scientist* 83: 148–57.

Flick, W. A. 1983. Observations on loons as predators on Brook Trout and as possible transmitters of infectious pancreatic necrosis. *North American Journal of Fisheries Management* 3: 95–96.

Fournier, F., W. H. Karasov, M. W. Meyer, and K. P. Kenow. 2002. Daily energy expenditures of free-ranging Common Loon (*Gavia immer*) chicks. *The Auk* 119: 1121–26.

Gill, F. B. 2007. *Ornithology*. 3rd ed. New York: W. H. Freeman.

Johnsgard, P. A. 1987. *Diving Birds of North America*. Lincoln: University of Nebraska Press.

McIntyre, J. W. 1988. *The Common Loon: Spirit of Northern Lakes*. Minneapolis: University of Minnesota Press.

Paruk, J. D. 1999. Behavioral ecology in breeding Common Loons (*Gavia immer*): Cooperation and compensation. Ph.D. diss., Idaho State University, Pocatello.

Strong, P. I. V., and L. Hunsicker. 1987. Sibling rivalry in Common Loon chicks. *Passenger Pigeon* 49 (3): 136–37.

Wright, K. G. 1998. Common Loon, *Gavia immer*, feeds on Pacific giant octopus, *Octopus dofleini*. *Canadian-Field Naturalist* 113: 522–23.

8

More Than a Foot Waggle

The Fascinating World of Loon Behavior

Over the course of the breeding season, loon behaviors are multifaceted, and as a researcher I have seen them behave in different or unusual ways. Behavioral ecology can sometimes enlighten us as to why this happens. One morning in late July 1993, two Earthwatch volunteers and I were watching a pair of loons and their two chicks on G Pool at Seney National Wildlife Refuge. We had been watching the family since 5:00 a.m., and our four-hour shift was almost up. Suddenly, a single loon flew over and landed on the lake. It was unbanded, and odds are it came from off the refuge since more than 80 percent of the adults at Seney are banded. Curiously, the parents left the chicks near an island and swam out to greet this new loon. The encounter appeared nonaggressive. The loons swam past each other in a line and then did some bill dipping. Next, they performed some circle swimming, followed by more line swimming. This interaction lasted for eleven minutes until another loon (also unbanded) landed on the lake at 8:58 a.m. This loon quickly joined the group, and then all four loons swam in a line, occasionally dipping their bills.

Overall, the interaction did not appear aggressive. There were no wails, tremolos, or yodels, for example, but I could hear some individuals hooting (a short-distance, nonaggressive vocalization). The territorial parents allowed this interaction to continue. They did not look back to see how their chicks were doing and seemed content to interact with these new loons. At 9:12 a.m., they all dove rapidly and surfaced for just a second or two, before plunging back beneath the water. The loons were no longer in a tight group; they were moving away from where they first dove into the water, toward the center of the lake. The four birds stayed in the center until 9:25 a.m., when a fifth loon (also unbanded) landed and quickly swam out to greet the other four. They greeted the new bird, and the five loons swam slowly, in a relaxed manner around each other away from the chicks. At 9:49 a.m., yet another loon landed on the lake, and then all six loons swam together, in stereotypical fashion. This continued for ten minutes.

Suddenly, a few individuals started jerk diving (the head is lurched forward in succession one or more times, before it arches back and then plunges into the water). Before long, the whole group was jerk diving and swimming in circles. This lasted for about one minute, and at 10:00 a.m., three of the loons reared up out of the water and performed an upright aggressive display. One loon let out a tremolo call, and several more joined in. Eleven minutes later, one of the loons swam away from the group on the surface, turned around, and faced into the wind. It then flapped its wings while gaining lift down the invisible runway and departed the lake. At 10:20 a.m., 10:24 a.m., and 10:25 a.m., the other loons too flew successfully off the lake, each departing into the wind. The original pair remained in the center for a few minutes before swimming back to the island where they had left the chicks, who swam out to meet their parents. *What was that all about?*

Social Gatherings

•

Starting in mid-July and continuing through August, loons take to the air and visit loons on other lakes. It is at this time of the year they are most conspicuous to boaters and homeowners on the water. What makes loons so noticeable is both their number and behavior. First, many loons aggregate, anywhere from five to twenty individuals (occasionally more), drawing attention to themselves by their sheer number. Then, their behavior, such as circle swimming and line swimming with one passing the others, is very noticeable. It is similar to hockey players shaking hands at the end of a game. I believe I can even make a case that loons suffer from a split-personality disorder—a kind of Jekyll and Hyde affliction. A stretch, you say? Perhaps, but here is my evidence for the analogy. In May, loons are highly territorial and asocial; they will attack and occasionally kill other loons in their territory. But come mid-July and into August, the same loon that aggressively attacked intruders two months earlier now tolerates their presence within its territory, despite having two chicks. Incredible, right? This aspect of loon behavior has always fascinated me, and I must concede that I am not alone, for I get more questions about it than any other.

Why do loons engage in this behavior? What is its function? Is it adaptive? Social gatherings occur at specific times and places and consist of highly stereotyped and ritualized behaviors (jerk diving, bill dipping, line swimming). Judy McIntyre's research has suggested that social gatherings serve to foster familiarity and cooperation among loons, which may be important during migration when large aggregations called feeding flocks can occur on big lakes. Being a member of a feeding flock has several advantages over foraging alone, for example, having a higher probability of locating and catching food. Loons flock in early to late spring, and again in late summer to early fall, but these aggregations often lack the distinctive ritualized and stereotyped behaviors. Social gatherings that occur earlier in the summer than fall flocking may simply be a transition time between degrees of sociality. Since loons may experience greater foraging success when concentrating on schooling fish (e.g., herring) versus solitary fish (e.g., perch), it would be to their advantage to tolerate and possibly even cooperate with other flock members. A loon that does not cooperate and hunts alone may have lower foraging success.

Walter Piper and I called Judy McIntyre's ideas on social gatherings the "familiarity hypothesis" simply because we needed to name them. For the most part, it appears to accurately describe them, but we wondered if there was an added layer of intrigue to these gatherings. Maybe loons were searching for new or unoccupied territories. If this were the case, then a large number of participants in social gatherings should be nonbreeders, and the social gatherings should occur on territories (or lakes) of loons in a nonrandom fashion. Our reasoning was that an unpaired loon without a lake or loon territory might use this strategy to acquire one. Also, not all loon pairs, territories, and lakes are alike. What if a current pair experienced a failed nest or brood? They might be less committed to each other and open to a new mate. We called this the "reconnaissance hypothesis." We also wondered if social gatherings were more likely to occur on territories that had chicks than those without because those territories would appear to other loons to be high quality (presence of chicks = high-quality habitat).

In an attempt to determine the function of social gatherings, I studied two populations of loons in Upper Michigan (Seney) and northern Wisconsin (Turtle Flambeau Flowage). At the time, the adult loon population at Seney was over 80 percent banded, and on the Flowage it was close to 60 percent. For the banded birds, I could obtain specific details, such as what lake or territory a bird occupied, its sex, and its current breeding status (e.g., paired or unpaired, with or without young). From July 15 to August 30 our team of Earthwatch volunteers and staff, often equipped with walkie-talkies, watched many lakes from the shore and gathered the following data on social gatherings: lake, territory name, location, beginning and ending time, number involved, whether individuals were banded or unbanded (if banded we recorded the color combination, if possible), a description of the nature of the gathering (were there aggressive elements, for example, yodel, tremolo, upright defense postures), and lastly, the weather (primarily wind). I also filmed, from start to finish, 31 social gatherings so I could better dissect the behavior I was observing. Over the four-year study, I gathered data on 320 social gatherings. This is what I found out.

Approximately two-thirds of all social gatherings occurred within the first four hours after sunrise, and one-third occurred during late afternoon or early evening. Social gatherings were not observed between 11:30 a.m. and 3:00 p.m. Gatherings were uncommon when winds were strong (greater than 10 miles per hour). Mornings were preferred to evenings because they tended to be less windy. The fact that the timing of social gatherings is predominately morning and late afternoon and evening suggests that selection is responsible for this pattern. Because less wind equates with calmer water and higher visibility, conditions that create optimal viewing from a distance, loons have a higher probability of spotting each other in the morning than the afternoon, when the winds typically pick up. Also, gathering at the same times, more or less, each day, allows for a highly efficient communication with other loons (like when humans go to the local pub after work for social interaction). But why partake in such behavior in the first place? What are these loons communicating?

Many birds form flocks in the fall, and such behavior provides several advantages, including an increased likelihood of finding food and a reduced risk of predation. We know loons flock in the fall, typically on large lakes, a behavior that has likely been shaped by selection. In addition, it appears that social gatherings, which take place from mid-July through August, may also have been influenced by selection. Some individuals may benefit from these interactions with other loons prior to fall migration. While measuring or quantifying the benefit is difficult, we can speculate that these gathering aid in adjusting hormone levels that bring about sociality. Perhaps without contact with other loons in late summer, loons would be more asocial during the fall. As McIntyre posited, social gatherings likely foster cooperation, which is important since nearly all loons join fall flocks. For example, in July 1995, my teams and I observed five loons (all banded) each morning at the Turtle Flambeau Flowage between 6:10 a.m. and 6:30 a.m. for three consecutive days. This group consisted of a female (successful nester) and two pairs of unsuccessful nesters from adjacent territories. Because cooperation is rewarded in group foraging, loons that tolerate and cooperate with other flock members will likely have better foraging success than those that don't. Thus, over time, selection may favor loons that cooperate in social gatherings.

In my observations, on a continuum from mid-July to the end of August, the number of loons engaging in social behavior increased as did the average time the group remained together. More aggressive displays such as tremolos, yodels, and upright defensive postures occurred in July than August. Social gatherings were also more likely to occur over deeper versus shallower water. At Seney, where 80 percent of the adult territorial loons were banded, slightly more than half of all participants in social gatherings were unbanded (55 percent), despite the great majority of all adults on the refuge being banded. Moreover, because the unbanded loons on the refuge (20 percent) rarely left their breeding lakes, we made the obvious conclusion: the unbanded loons we were observing were coming from outside the refuge. Did these unbanded loons hold territories on other lakes, or were they unpaired (floaters) and without a lake? How far did they travel to get to Seney National Wildlife Refuge?

Due to Dave Evers's determination, he and his teams have banded many loons within 20 miles of the refuge, often working through the night (and without funding). Yet, none of these loons were observed on the refuge. Were they coming from an even farther distance, or, again, were they nonbreeders on lakes in the area that we hadn't caught and banded? Dave's night-lighting technique works well to capture loons with young, but if a loon did not have young (had an unsuccessful nest or was unpaired), it often remained unbanded. Lake Michigan is 25 miles south of Seney, and 30 miles to its north is Lake Superior.

Nonbreeding loons are commonly observed on these lakes during the summer, so how long would it take for them to make a flight to Seney? Flying at a speed of 70 to 75 miles per hour, loons could reasonably reach Lake Michigan and Lake Superior in under thirty minutes. I cannot be certain where those unbanded loons at Seney originated, but I think some of them came from one of those Great Lakes. Why were these unbanded loons making the effort to fly to Seney?

Periods of post-breeding dispersal are common in birds, and it has been suggested this behavior may lead to acquiring territories or future mates. I documented three cases where a banded loon (failed nester) was observed at a social gathering on the same lake in successive years. Walter Piper observed a similar situation where one loon ended up nesting on the lake it had visited the previous year. Therefore, it is possible that loons are acquiring information about the status of the resident pair and the quality of the habitat, which suggests social gatherings may serve as a kind of information center. I found it curious that at Seney a significant increase in the number of social gatherings occurred on lakes where the territorial pair had broken up. Was it a mere coincidence that territories that incurred a breakup in year one also experienced a fair number of social gatherings in year two? Why would loons tend to aggregate in that territory and not on other lakes? Like other long-lived birds, such as crows, loons have long-term memory. My observation data suggest that loons are not aimlessly wandering among lakes but are instead gathering information about the status of the pair that has established the territory they are visiting. We found no significant difference in the number of social gatherings observed between successful territories (with chicks) and unsuccessful territories (without chicks). Piper's similar study, which monitored sixty lakes, found that social gatherings were more common on loon territories that produced chicks than on those that failed to produce them.

Of the banded loons, unsuccessful nesters were more likely to partake in a social gathering in July than successful nesters. However, unsuccessful and successful breeding loons participated equally in social gatherings during August. This suggests that one or both of the parents felt more comfortable leaving their chicks later during the breeding season, knowing they are fairly capable of eluding any potential threat. Older chicks are further along developmentally and can dive and elude predators better than younger chicks. Also, if leaving their young to participate in social gatherings led to higher chick mortality, then this behavior would be selected against, and individuals that remained with their young would be rewarded with higher chick survival. The fact that one or both adults leave to join in social gatherings suggests the benefit of doing so does outweigh the potential risk of losing their young, and that beyond a certain age, the probability of the chicks being preyed upon is very small. I would predict that the best time for a loon to safely leave its chicks alone is when they can swim independently, which happens around six weeks of age. Yet I have seen both adults leave when the chicks were only four weeks old. One thought about why they engage in this behavior is that they simply do not have a choice. If one or more loons land in a loon's territory, it must greet them aggressively, greet them cautiously, or not greet them at all. I think some adults swim out to meet the intruders and then lead them away from the young because that is their best option. Out of my 320 observations, we never had a case where a loon carried out infanticide.

In the film that I captured during this study, 95 percent of all the loons peered below the surface of the water at least once during a social gathering (loons peer to observe what is underwater). Some individual loons peered more than others, and some groups peered more frequently than other groups. We correlated the level of aggression with the amount of time spent peering (more peering = more aggression). We had difficulty confidently identifying the event that triggered the aggression. Did an individual approach too close? Was there some type of miscommunication or misinterpretation of a signal or behavior? During the breeding season, loons are suspicious of other loons. This conspecific suspicion is hardwired into their DNA. Given this behavioral trait increases their odds of reproductive success, it pays for loons to notice what other loons are doing. They take cues from other members in the flock, and if one loon in the group is peering, odds are there must be another loon underwater, and thus others will peer also.

In summary, Common Loons change their behaviors over the course of the breeding season. They can go from being strongly territorial (attacking other loons) to intermingling with the same individuals they vigorously confronted just two months prior. That both parents leave their young to partake in social gatherings in late summer suggests they gain some advantage for doing so. Social gatherings likely foster cooperation among potential flock members, and they may facilitate reduced levels of aggression, which could be important for the formation of foraging flocks commonly observed during migration. Over time, loons may use these opportunities to gather information about the availability of a mate or a territory. I think both the familiarity and the reconnaissance hypotheses suggest valid explanations for social gatherings, and we should recognize that they serve as important information centers for breeding loons.

Foot Waggle

•

Many observers have noticed that loons often roll over on one side, lift half of their body and one leg out of the water, stretch the leg, and then shake it quite vigorously; hence, the name foot waggle. All loons foot waggle, even chicks. Why perform this behavior? Diving ducks, grebes, and mergansers also foot waggle, so it is not unique to loons. Frank McKinney, distinguished professor at the University of Minnesota, wrote the treatise on comfort movements and foot waggling in ducks. He noticed that ducks take one leg out of the water and rest it on their back, but before "shipping" it, they shake it to remove dirt and debris they may have picked up while probing in the substrate so as not to transfer it to the skin or feathers. Birds about to sleep or rest often perform a foot waggle before they ship their foot, which supports McKinney's theory that it is primarily a comfort movement.

Loons have extraordinarily big feet, and many birds dissipate heat through their legs. Judy McIntyre proposed that loons use their feet for thermoregulation, but could the simple act of a foot waggle lower or raise the core temperature? And if so, how? From dawn to dusk I gathered observational data for more than four thousand hours on twenty-two loons (eleven pairs). For each foot waggle, I took note of several things: the sex of the loon, time of day, frequency, whether the waggle occurred on the right or left leg, what type of behavior preceded the waggle, and what occurred during and after each event (preening, sleeping). I also recorded the following four environmental variables hourly: air temperature, amount of incoming solar radiation, wind speed, and water temperature. In addition, I kept track of foot waggles when a Bald Eagle flew overhead or landed nearby. In an earlier pilot study, I noticed that some loons foot waggled while interacting with other loons during social situations in late summer. This struck me as odd. The observations led me to hypothesize that loons might foot waggle when they are anxious. Similarly, I kept track of when a chick foot waggled and in what situations and context.

I found that the loon foot waggle is primary a comfort movement but may also occur during potentially anxious situations. I did not find a mechanism for its use in thermoregulation. A loon foot waggles prior to shipping its foot to remove dirt or debris, and first thing every morning (84 percent of all adult loons that I observed would perform one during the first hour of the day). In the first case, the foot waggle is done to prevent contamination of the skin and feathers, and in the second one, to simply stretch the leg after resting for six to seven hours. I found some evidence that loons foot waggle in potentially anxious situations. For example, nearly a third of all loons foot waggled (usually just once) when approached by a speeding boat or when an eagle flew into a loon territory. I also observed at least one loon foot waggle during a social gathering nearly half the time. In those situations, loons may be expressing anxiety or discomfort just as they may vocalize or swim away for the same reason.

But what about the foot waggle being involved in thermoregulation? First, let us examine a foot waggle's possible role in heat gain, followed by heat loss. For something to gain heat, it needs a large surface. A loon foot would appear to have that, but large is relative. The body of a loon is at least ten times larger than its foot. So why not simply use the body? For the solar collector to do its job, it needs to remain exposed for a long duration, not just a few seconds. Think of how solar panels are exposed to the sun all day. The duration of a foot waggle is simply too short, a matter of seconds, to absorb any heat. On occasion, a loon will open its foot and stretch the webbing between its toes, creating a larger surface, but again the duration is too short to absorb heat. But maybe the foot waggle promotes heat loss. If a loon is overheating, the fastest way to cool down is to swim underwater, as water conducts heat away thirty times faster than air. If they cannot or choose not to dive underwater, they could increase blood flow to their legs and extend their feet and spread their toes. Because water is a better conductor of heat than air, shaking the foot in the air is not the most effective way to lose heat. To be clear, the loon foot is absolutely involved in thermoregulation, but the act of a foot waggle has little to do with heat gain or heat loss.

Nevertheless, I did obtain some interesting results by measuring the frequency of adult foot waggles and some environmental variables. Although the number of foot waggles was not associated with ambient temperature or water temperature, waggles increased 10 to 12 percent on sunny days compared to cloudy, and decreased 10 to 12 percent on windy days. Why would a loon foot waggle more on a sunny day? My data showed that prior to nesting, adult loons foot waggled eight times a day, and while nesting, seven times a day, but during chick rearing, they foot waggled thirty-seven times a day. Adults spent on average 158 minutes per day resting on the open water prior to incubation, and 58 minutes per day while incubating (most of their time when not incubating was spent foraging). When they had chicks, they spent 269 minutes per day (4.5 hours) resting on the water, or almost 2 hours more a day absorbing sunlight. The amount of solar radiation can have a dramatic effect on a bird's heat load. Are adult loons experiencing heat stress during mid to late summer while resting with their young? Was this another example of a stressful situation and a loon responding to it by foot waggling? I suspect so. While they foot waggled both right and left legs equally, females performed this behavior more than males, on average forty-three to thirty-three times per day during the chick-rearing period. Why this is so remains unclear. I suspect that the loons foot waggled less on windy days because they could not afford the luxury of shipping a foot—they needed both feet to swim and maintain their position in the water.

All four environmental factors (air temperature, solar radiation, wind speed, and water temperature) were significantly associated with foot waggling in chicks. Like adults, chicks reduced their foot waggles on windy days and increased the number on sunny days. However, both increasing ambient and water temperatures led to more foot waggles. Later in the summer chicks are at their largest, while water (and to a lesser extent air) temperature is at its warmest. I think the positive correlation exists because an older chick is absorbing more heat and losing less heat to its thermal environment than a younger chick. An older chick is experiencing more thermal stress, and, like the adults, it also foot waggles. On average, chicks foot waggled forty-four times per day. Also, some associative learning may have been going on between the chicks and the adults, because many times a chick would foot waggle after watching an adult perform the behavior.

Nocturnal Behavior

•

For a long time, I wanted to observe loons at night. Who wouldn't want to sit in a canoe or nestle on shore, and look out over the open water and up at the sky, relishing the solitude and tranquility. And then there's the unknown: what do loons do all night? Loons vocalize at night so in addition to sleeping, so they must be doing something. Do they patrol their territories? Get into contests with other loons? With Earthwatch funding I was able to purchase a pricey pair of night-vision binoculars and seize the opportunity to find out for myself. The plan was to observe several pairs on different sized lakes. Medium-sized lakes could support only a single pair, and the larger lakes could support multiple pairs. I speculated that loons occupying larger lakes likely patrolled more and had more interactions with neighboring pairs than those on medium-sized lakes. We worked in two shifts, one person observing from 10 p.m. until 1:00 a.m., and the other from 1:00 a.m. until 4:15 a.m. Time for a confession: I was so busy running the daytime program that I did not have the time to do many night shifts. My assistants did the heavy lifting on that research and left with far richer memories (though one night I did observe a deer swim across a shallow section of the lake).

Overall, we did not observe any feeding. The loons preened on occasion, but they mostly rested in protected coves or out in the middle of the lake. They slept with their heads tucked and bills nestled under their back feathers. The average sleeping bout lasted twenty-four minutes, but it varied from four to fifty-four minutes. Overall, adults averaged about 1.5 hours of sleep a night. Nearly all their time was spent resting and sleeping (more than 90 percent). When we compared lakes, we found that loons on the large lakes spent more time surface swimming and patrolling their boundaries than loons on the medium-sized lakes. We also observed two territorial skirmishes between loon pairs on the large lake. At least one adult, and sometimes both, was always with the young.

Behavioral ecology has provided important insight into loon behavior, but there are still unknowns and areas to pursue. For example, are certain loons more bold or aggressive than others in certain situations and contexts? Is the same bold individual always bold across similar, but different, situations? If a loon is bold toward other loons, is it also bold toward non-loons, such as ducks or people intruding in its territory? Animals, like people, have personalities, and those personality traits are passed down generationally. If boldness benefits the individual and if this behavior is inherited (has a genetic component), then it would be selected and potentially passed down to future generations. But what if boldness is not advantageous in all situations? If a situation or context required a loon to be less bold, selection would swing the pendulum in that direction. For example, perhaps when loons are around humans, being less bold or conspicuous is beneficial. The point is that a species has a range of behavioral reactions to situations and contexts, which likely benefits them as it may promote flexibility and adaptability. Further work on suites of correlated behaviors in loons should yield new insights into their behavior and their potential to adapt to a changing landscape.

Further Reading

Barr, J. F. 1973. Feeding biology of the Common Loon (*Gavia immer*) in oligotrophic lakes of the Canadian Shield. Ph.D. diss., University of Guelph, Ontario.

Barr, J. F. 1996. Aspects of Common Loon (*Gavia immer*) feeding biology on its breeding ground. *Hydrobiologia* 321: 119–44.

Gostomski, T. J., and D. C. Evers. 1998. Time-activity budget for Common Loon, *Gavia immer* nesting on Lake Superior. *Canadian Field-Naturalist* 112 (2): 191–97.

McIntyre, J. W. 1988. *The Common Loon: Spirit of Northern Lakes.* Minneapolis: University of Minnesota Press.

McKinney, F. 1965. Comfort movements of the Anatidae. *Animal Behavior* 14: 178–205.

Paruk, J. D. 1999. Behavioral ecology in breeding Common Loons (*Gavia immer*): Cooperation and compensation. Ph.D. diss., Idaho State University, Pocatello.

Paruk, J. D. 2006. Testing hypotheses of social gatherings of Common Loons (*Gavia immer*). *Hydrobiologia* 567: 237–45.

Paruk, J. D. 2009. Function of the Common Loon foot waggle. *Wilson Journal of Ornithology* 121: 292–98.

Paruk, J. D. 2009. Nocturnal behavior of the Common Loon. *Canadian Field Naturalist* 122: 46–49.

Piper, W. H., J. D. Paruk, D. C. Evers, M. W. Meyer, K. B. Tischler, M. Klich, and J. J. Hartigan. 1997. Local movements of color-marked Common Loons. *Journal of Wildlife Management* 61: 1253–61.

Piper, W. H., K. B. Tischler, and M. Klich. 2000. Territory acquisition in loons: The importance of take-over. *Animal Behaviour* 59: 385–94.

Piper, W. H., C. Walcott, J. N. Mager III, M. Perala, K. B. Tischler, E. Harrington, A. J. Turcotte, M. Schwabenlander, and N. Banfield. 2006. Prospecting in a solitary breeder: Chick production elicits territorial intrusions in Common Loons. *Behavioral Ecology* 17 (6): 881–88.

Sih, A., A. Bell, and J. C. Johnson. 2004. Behavioral syndromes: An ecological and evolutionary overview. *Trends in Ecology and Evolution* 19 (7): 372–78.

Wentz, L. E. 1990. Aspects of nocturnal vocal behavior of the Common Loon. Ph.D. diss., Ohio State University, Columbus.

9

Loons on the Move

THE STRATEGY AND DYNAMICS OF MIGRATION

Wildlife research involves roving parts, and sometimes things do not always go according to plan. Here's a good example. In 1998, Dave Evers and I were at Walker Lake, Nevada, teaming up with researchers Mike Yates and Mark Fuller from Boise State University, Larry Neel from the Nevada Department of Wildlife (NDW), and Kevin Kenow from the U.S. Geological Survey (USGS). Each team was there to do a specific job: Mike and Mark from Boise State were the leaders of the project; the team from NDW had the boats, drivers, and expertise to navigate around the lake, and the USGS staff were responsible for conducting surgeries and implanting satellite transmitters. Dave and I were there to catch the loons. We all met the third week in April.

During the 1990s, Larry Neel began noticing hundreds and on some days more than a thousand loons on Walker Lake during fall and spring migration. It was suspected that these loons bred in Canada and likely wintered off the California coast. Walker Lake is a high-elevation lake, 11 miles long by 5 miles wide, located 75 miles southeast of Reno and surrounded by the Wassuk Range. The Walker River drains into Walker Lake, which has no outlets (called a terminal lake). Thousands of years ago, this area was covered by a much larger waterbody, Lake Lahontan, but over time it shrank, and what remains today is Walker Lake. Loons stop at the lake because fish of the right size are prevalent. The lake supports a large native population of tui chub and Lahontan cutthroat trout. The local town of Hawthorne has an annual loon festival during the peak of the migration.

We went through introductions and discussed the plan for the next few days. The Boise State team notified the group that the satellite transmitters were not in hand but would arrive the next day, so we questioned whether to put off going in the boats until then. Strong winds were forecasted each evening, and going out on the lake was potentially too dangerous. The winds from this region are known to howl down the mountain and through the valley, creating large whitecaps and dangerous conditions. We tossed around our limited options. Dave and I offered that we were not sure we could catch loons outside of the breeding season. Recall that, on breeding lakes, adult loons with chicks will remain on the surface of the water and are fairly easy to catch, but without chicks around, their parental instinct to protect their young does not exist so there is nothing to prevent them from simply diving each time they see the light. Because we did not know if we were going to have success or how strong the winds would howl each night, we decided to go out that night and at least try. We thought that maybe we might learn some little trick that might help us the next night.

We took three boats out. Dave, an NDW biologist, and I were the netters in each boat. The boat captains from NDW were extremely competent, had GPS navigation on each boat, and were eager to get underway. We drove around the lake with a spotter shining the light out over the water. Dave and I used binoculars and tracked the beam of light, looking for the silhouette of a loon. On clear nights one can spot a loon 100 yards away. At first, we noticed only many grebes and waterfowl but eventually came across a few loons. Waves of 1 to 2 feet accompanied us on the water, which worked to our advantage because they broke up the noise of the boat engines and limited the range at which a loon could spot us. Thus, when a loon came into view, we were already pretty close to it, giving it less time to react.

The hard work continued all night. When we did locate loons, several of them dove, forcing us to search elsewhere. But a few paused upon seeing the light, and this gave us a chance to net them. We caught five loons that night, but we were faced with the hard reality that the satellite transmitters wouldn't arrive until the next day, and we had no place to keep the loons overnight. Did we release them and try our luck tomorrow? What if the weather conditions worsened, as expected, and we could not launch the boats? Loons are big, hardy birds, and with our collective experience we felt the loons would be fine if we kept them overnight. But where? With limited options, we reasoned we had to take them with us to our hotel rooms. Was this going to work? Seriously, did we tell the hotel manager we would each have an additional guest in our rooms?

We decided to keep the three largest individuals for the surgeries. Kevin injected saline solution under the skin to keep them hydrated, put them in cardboard boxes (with air holes), and placed a life jacket in the bottom of the box to protect the birds' keels. We placed one loon per room and turned on the air conditioners to keep them cool. The loons were calm, but we discussed that the loons might be better off if their weight was supported by water. I forget whose idea it was to fill up the bathtub, but we did, and Dave lowered one of the loons into it. The bird did not panic (thank goodness) and simply floated there. We posted one person in the bathroom in each room to make sure the loons were safe. Most of us were sleep deprived. Morning could not come fast enough, but our wait was extended, because the satellite transmitters did not arrive until 12:30 p.m. We drove the loons to the local volunteer fire station nearby. Kevin had sanitized the building and prepared it for the operations. I appreciated his calm demeanor and professionalism and was more than glad to give the loons to him.

After they were sedated, Kevin made a small incision along the back between the scapulars and placed the transmitter beneath it. He then sewed up the incision with a couple of stitches. I must confess I found it odd to see a loon under anesthesia, its body limp and stretched out over a table, with an oxygen regulator nearby and everyone wearing face masks. The surgeries went well, and we later released the loons at the lake. We watched them over the next two days, and within a couple of weeks they flew off the lake. Mike Yates tracked them as they migrated and gave us updates (one of the loons flew to Saskatchewan). It's true: things do not always go according to plan, especially in wildlife research. None of us liked the idea of holding those loons for more than ten hours, but given the information we had, we decided it was best to hang onto them for the night rather than release them. That decision turned out to be the right one because the other nights during our stay were so windy we never did get back on the water. In 2000 we went back to Walker Lake, caught two more loons, and implanted satellite transmitters in them. Both birds flew to Saskatchewan and remained there for the summer.

The Timing of Migration

•

In North America, many songbirds migrate south in the fall because their steady supply of insects is no longer available. Since they have wings, songbirds can just pick up and fly to a more suitable location. Other birds, such as grebes and loons, migrate because ponds and lakes freeze over, rendering their food inaccessible. They have no choice. Migrate and live, remain and die. Nearly all loons choose the former, but each year a few immature loons remain on some northern lakes into December. This is not a wise choice. Many will get iced in and may fall victim to a Bald Eagle or a coyote. Over millennia, selection has operated on the timing of loon migration to maximize their fitness (survival and reproductive success). Staying put, with the risk of the lake freezing over, is not a winning strategy. Loons require a long runway to take off, much like an aircraft, and expanding lake ice decreases its length. If a loon cannot reach a critical speed at takeoff, it cannot overcome gravity and attain lift. The great majority of loons make the right decision in the fall. They leave before the ice settles in around them, and those that stay fail to contribute their genes to the next generation.

Fall Preparation and Departure

By the end of August, most loon chicks are competent swimmers and adept at catching their own food (assuming they hatched in early July), but depending on their age, they may require some supplemental food from one or both parents for a few more weeks. Loons with young stay longer on their breeding lakes because they have parental responsibilities. If loons do not produce young or if their young die, they have no parental responsibilities and are free to remain on their breeding lake or migrate to another lake or the coast, depending on their proximity. In our Biodiversity Research Institute pilot migration study in Maine (discussed in chapter 4), we found that males departed three to six weeks sooner than females. In that study, we had only six individuals, too small a sample size to draw any strong conclusions. Kevin Kenow compared departure dates from twenty-one pairs of loons breeding in Minnesota and Wisconsin, and he found no difference in departure dates between the sexes. From that study, it appears that sometimes males leave before their partners, and at other times the females leave first, and thus selection has balanced departure dates between pair

John James Audubon, *Great Northern Diver, or Loon,* 1836.
Bell Museum Collection, University of Minnesota.

•

A mated pair of Common Loons in breeding plumage swims on a lake in Yellowstone National Park. Photograph by Charlie Hamilton James/National Geographic.

Red-throated Loon *(top left)*, Pacific Loon *(top right)*, Yellow-billed Loon *(bottom left)*; photographs by Ryan Askren, U.S. Geological Survey. Arctic Loon *(bottom right)*; photograph copyright Jari Peltomäki/Finnature.

A breeding adult Common Loon performs a wing flap, which helps to realign the feathers. Photograph copyright Jeff Bucklew.

•

Because of their size, adult loons need a long runway to build up the speed required to achieve lift. Photograph copyright Jeff Bucklew.

Confrontations between loons can be intense. Here two loons fight over a territorial dispute. Photograph copyright Daniel Poleschook and Ginger Gumm.

•

During the breeding season, when loons feel threatened (such as when people get too close to the nest), they display a distinctive penguin-like pose in which they rise out of the water and face the intruder. Photograph copyright Daniel Poleschook and Ginger Gumm.

The deep, clear lakes in British Columbia provide ideal nesting habitat for Common Loons.
Photograph copyright Roberta Olenick, neverspook.com.

•

Because of their proximity to the shore, loon nests are vulnerable to fluctuating water levels.
Photograph copyright Daniel Poleschook and Ginger Gumm.

Nesting loons carefully tend to their eggs. Here a recently hatched chick vocalizes next to its parent. Photograph copyright Roberta Olenick, neverspook.com.

•

Parents cooperate in feeding their chicks. This young chick receives a leech from one of its parents. Photograph copyright Roberta Olenick, neverspook.com.

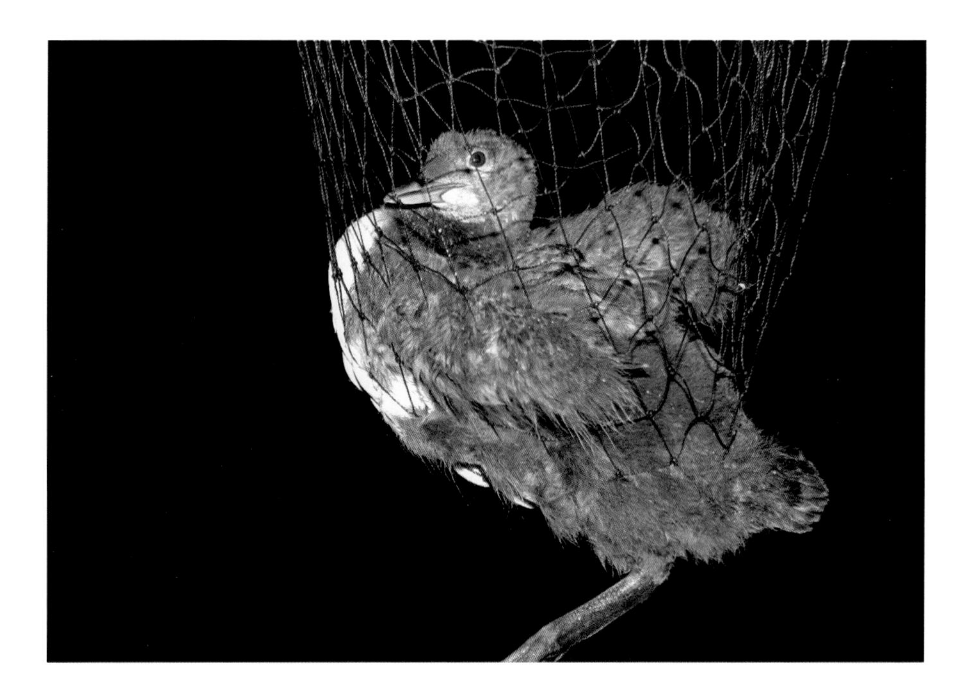

A mature loon chick is hoisted from the water during a night capture expedition.
Photograph copyright Daniel Poleschook and Ginger Gumm.

•

The tremolo is one of several loon vocalizations frequently heard by lakeshore visitors.
Photograph copyright Daniel Poleschook and Ginger Gumm.

While preening the feathers on its belly, a loon will sometimes extend a leg while its foot waggles. Photograph copyright Jeff Bucklew.

•

In late summer, loons often congregate in groups, called social gatherings, such as this one on a lake in Maine. Photograph copyright Paul Tessier/Stocksy United.

A wintering loon begins to molt. The breeding plumage is still intact on much of the wings. Photograph by Darwin Long IV.

•

An adult loon in full winter plumage. Note the squared tips of individual feathers. Photograph copyright Daniel Poleschook and Ginger Gumm.

A juvenile loon in winter plumage. Note the rounded tips of individual feathers. Photograph by Darwin Long IV.

•

Crabs are common in the diet of wintering loons. Photograph by Steve Byland/Shutterstock.

Lead poisoning from ingested fishing tackle is a serious problem for loons in both breeding and wintering habitats. This loon died after ingesting lead fishing tackle on a lake in Washington. Photograph copyright Daniel Poleschook and Ginger Gumm.

•

Discarded fishing line left by anglers is one of several environmental threats to loon populations. Photograph copyright Roberta Olenick, neverspook.com.

In recent years nesting loons have been greatly aided by the implementation of artificial rafts with avian guards, which protect them from overhead predators. Photograph by Jonathan Fiely.

•

A relocated loon acclimates to a holding pen before being released on a Minnesota lake. Photograph by Michelle Kneeland.

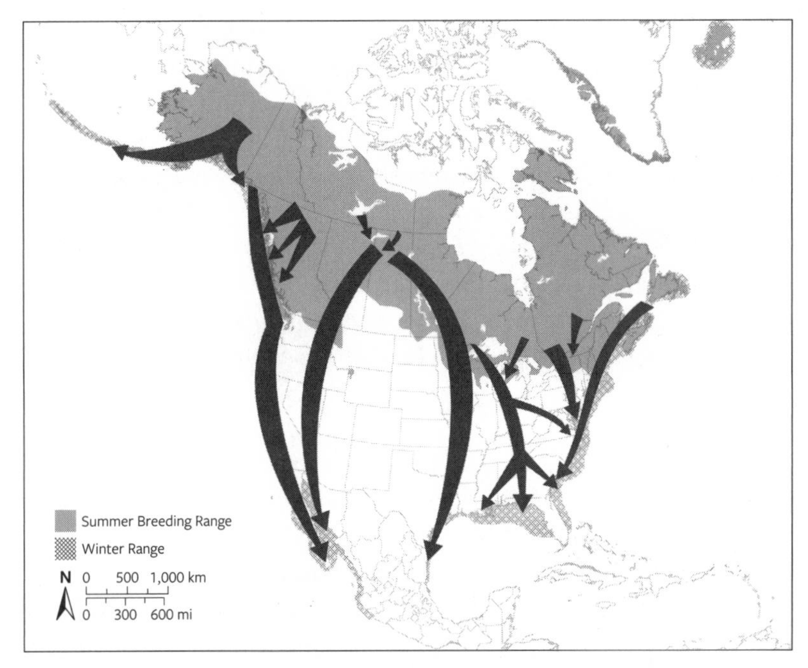

Common Loon fall migration routes. Map by Mark Burton, Biodiversity Research Institute.

members. But it would be interesting to know what factors are involved in shaping which loon leaves first and why. Is it age? Individual health? Genetic? We still have much to learn about these birds.

In our Maine study, the males flew directly to the ocean because they could easily cover that distance in two to three hours. Instead of flying due east to reach the ocean, they flew northeast, to the Bay of Fundy between Nova Scotia and New Brunswick. The Bay of Fundy is one of the most food-rich places on the Atlantic Seaboard and supports a predictable, abundant fish population. In addition to loons, many other migrating birds, such as sandpipers and plovers, use this area for extended periods to replenish their fuel (fat) stores. The loons we tracked stayed in the Bay of Fundy for three to six weeks, a finding that supports its value as a staging area. This pattern of leaving the breeding lake and stopping at an annual staging area suggests that food is the driving force behind these loons' movements (even a couple of Kenow's loons from Minnesota flew there!). For loons breeding in the interior of the continent, the ocean is a long way away, and better options may be, for example, the Great Lakes or other large interior lakes, such as Mille Lacs, Red, and Winnibigoshish in Minnesota and Lake Winnebago in Wisconsin. Here they remain for several weeks, much like loons visiting the Bay of Fundy, because these lakes have a fishery that can support hundreds of loons. In many instances, these lakes have an abundance of schooling fish, such as herring, alewives, and shad, which can be hunted more effectively in groups rather than by individuals. This is why so many loons switch from solitary foraging in the summer months to cooperative foraging in the fall and winter. By working cooperatively, loons gain weight in preparation for their long migration to the coast.

Hyperphagia

Migrating birds exhibit a kind of fall restlessness. German ornithologists call this *Zugunruhe,* and it entails excessive eating, or hyperphagia. Sandpipers, such as Red Knots, yellowlegs, and Sanderlings, are known for prodigious consumption of crustaceans during fall migration. Loons also eat a lot during the fall and consequently put on fat. Loons likely undergo physiological changes that alter their metabolism so that the production of fat increases. A gram of fat can store twice as much energy (when burned) compared to a gram of protein or carbohydrate, so it makes sense that many migrant birds use this approach. Loons store fat throughout their body and use it for fuel during migration. By late fall, a loon is ready to migrate to its overwintering location. But on what specific day or under what environmental conditions will an individual loon depart?

When to Leave (for Good)?

Many loons migrate in stages, but environmental factors such as wind direction and air pressure also shape departure dates. Conditions on some days are better for flying than others. For a loon during fall migration, that means flying with a tailwind (out of the north or west) rather than into a strong headwind (out of the south or east). Loons that perceive and use wind direction as a selection criterion will save energy during their flight and arrive to their wintering grounds in better condition than loons that spend more energy flying into the wind. Selection also influences the time of day a loon departs. A loon that makes the better choice will arrive in better condition than a loon that makes a less optimal choice. I have spent some time at Whitefish Point, in

Michigan, watching loons migrate both in the spring and fall, and my observations suggest that they leave in the morning, never in the afternoon. The reasoning here is that they have plenty of daylight to reach a suitable waterbody on which to land before nightfall. While migrating, loons may join up with other individuals to form a small group, a loose aggregate of three to fifteen individuals, but they rarely fly as a tightly knit flock or in V-shaped formation.

How High and How Fast?

•

During migration, birds fly at different elevations, mostly to maximize travel distance by taking advantage of both less turbulent and less dense air in higher elevations. Some birds will fly at 1,000 feet above ground level, others at 10,000 feet. Inland loons fly between 2,700 and 6,200 feet above ground level (considerably lower along the coast). Our Darwinian approach suggests that this elevation range is most likely optimal for loon migrants. Flying at higher elevations might take too much energy without a compensatory return, and there may be too much air turbulence at lower elevations. Unlike raptors, loons rarely use warm, rising air currents, or thermals, during the day, but a few individuals may (they have been observed gaining more than 2,000 feet in three minutes). Another likely benefit for loons flying at higher altitudes is cooler temperatures. Rapidly beating wings generate lots of heat. Because the additional heat load can cause physiological stress, it is advantageous for a migrating loon to fly where the air is several degrees cooler than at ground level. Air cools at roughly 5°C for every 1,000 meters (about 3,330 feet). Thus, by flying a half mile or more above the ground, a loon can increase its endurance through delaying overheating and minimizing the physiological adjustments necessary for cooling down.

A bird's wing shape and mass determine its optimal speed of travel. Flying slower or faster than the optimal speed results in less efficiency and greater effort. Loons, with their narrow wings and heavy body mass, lose elevation on the recovery stroke if they fly too slowly. To prevent this, loons beat their wings rapidly (more than 200 beats/minute), which increases flight speed. Consequently, the optimal flight speed for loons is a fast 70 to 75 miles per hour, compared to 40 to 45 miles per hour for smaller, aerodynamic ducks. A slower speed would be less efficient, and the cost of transport would be considerably higher. Although this may seem counterintuitive, it is most efficient for a migrating loon to beat its wings rapidly and maintain a high speed of travel.

Wing Loading and Migration Efficiency

•

The type of environment a bird chooses to live in, as well as its way of exploring and using its habitat, is closely related to its body size and wing shape. Migration distance is an outcome of the space that separates breeding and wintering habitat. In some birds that distance is relatively small, but in others it can be considerable. The greater the distance between breeding and wintering habitat, the greater strain, risks, and energy costs. An important factor that influences a bird's flying ability and migration distance is wing loading. To calculate it, one divides the bird's weight by the area of its wings. The shape of a bird's wing can have a great influence on its wing loading. For example, long and narrow wings provide less surface area to attain lift compared to long and broad wings. Birds with higher wing loading, such as geese, swans, and loons, have to work harder to stay aloft.

A loon's wings have been shaped by selection (see chapter 3) to minimize resistance and drag in the water during a dive, but this comes at a cost. Though long, a loon's wing is very narrow for a bird its size, and compared to other birds', its surface area is small relative to its body weight. This means that the load on the wing is high, and it takes more energy for a loon to fly compared to a duck, for example. In fact, loons have some of the highest wing loading values of all North American birds. Any discussion on migration in loons has to keep this principle in mind. Dave Evers and I began taking wing measurements of loons in the field in 1997. Using large rolls of drawing paper and a lightweight folding table, we would raise the loon to the height of the table, extend the wing over the drawing paper, and trace the outline. We recorded the circumference and the weight of the loon as well. In later years, we no longer needed to trace the wing but simply extended it, took a picture, and used a computer to calculate its surface area. We now have wing loading measurements from 143 loons across North America.

What we found is that, first, weights of North American loons vary greatly. For example, Common Loons breeding in the interior of the continent (Minnesota, Manitoba) are smaller (3,500 to 4,500 grams) and weigh less than loons breeding closer to the coast (such as in New Hampshire or Maine) (6,000 to 7,600 grams). Females are also 20 to 25 percent lighter than males for any given area. Why are loons breeding closer to the coast larger than those breeding in the interior? Loons that breed farther inland have a considerably longer journey to their ocean wintering grounds. The journey can sometimes be more than 1,000 miles, versus 250 miles for a loon closer to the coast, and the efficiency of flight depends on body size and wing shape. A smaller body size would likely be favored in loons that perform long-distance migrations in order to minimize energy expenditure and reduce migration time. Researchers in Europe examining migration distance in swans, for example, have observed that the smallest species (Bewick's Swan) had the longest migration distance, the largest species (Mute Swan) had the shortest distance, and the middle species (Tundra Swan) migrated an intermediate distance between the other two.

Second, a loon's wing shape also varies by population but only in males. There was no difference in wing loading values between males and females from either short- or long-distance migration populations. Since body size (volume) is a cubed function (length × width × height), and surface area of the wing is a squared function (length × width), any change in body mass (volume) will have a greater influence on wing loading than any change to the wing. For example, if we double both the volume and wing area of a loon, the volume increases eightfold ($2^3 = 8$), but the surface area increases only fourfold ($2^2 = 4$). So the fastest way for a migrating loon to save energy while flying would be to lose weight (i.e., get smaller), and for long-distance male loons, this has in fact happened.

Individual body size among animals is a central characteristic on which several life history traits may depend, including survival and number of offspring produced. In species with broad geographical distribution (like loons), body size may differ between populations because of local genetic modifications that could be a result of adaptation. If body size is heritable, then natural selection may operate to change average body size within species if there are direct fitness benefits. Migratory species make annual movements between seasonal environments, and the optimal management of time would serve to enhance their fitness. Natural selection rewards loons that are most efficient during migration, and the smaller body size of male loons that undergo long-distance migration illustrates this principle.

Patterns of Distance and Direction

•

Maintaining a speed of 70 to 75 miles per hour, a loon can fly between 560 and 600 miles a day if it flies for eight hours. This means that a loon breeding near the East or West Coast can make it to the ocean in half a day or less (about two to four hours), a leisurely journey by any standards. However, a loon breeding in the interior of the continent, such as in Minnesota or Manitoba, will have a considerably longer journey, measured in days, not hours. What is the migration strategy for a loon breeding farther inland? Does it power out consecutive long days of flight, or take shorter sojourns?

The spring migration strategy for a male loon we caught in Louisiana in 2011 consisted of a series of two-day flights followed by a period of staying put for several days (a range of four to fifteen days). Each two-day migration movement was around 600 to 675 miles, but in each case, the flight on the first day was considerably longer than the second day. Why did the loons stay put for one to sometimes two weeks? For example, a male we captured in Louisiana stopped at the J. Percy Priest Reservoir in Tennessee (surface area of 22 square miles) for nine days, and at Lake Michigan for fifteen days after two consecutive flights. On May 7, 2011, a female loon departed Buckhorn Reservoir in North Carolina and did not stop until it reached the Patuxent River in Maryland, a distance of 700 miles. The loon remained on the reservoir for eight days before departing on May 15. The data suggest that these loons required time to recoup the energy lost on their flights and were likely building up fat reserves for their next extended push northward.

The patterns I observed during the 2011 spring migration did not hold for fall migration, which was a much more leisurely affair. There is a sense of urgency in the spring for loons to get back to the breeding lakes, find a mate, defend a territory, and reproduce. In the fall, there are no breeding constraints, just the inevitable ice formation around November or December, and this gives them a bit more time to linger if they find a good feeding area. For example, in the spring it took two loons banded in the Gulf of Mexico 36 days to reach Saskatchewan but

100 to 126 days (3.5 to 4 months) to return to their winter location in the fall.

Natural selection shapes loon migration patterns. Loons that make suboptimal decisions will have lower survival rates and lower reproductive success than individuals who make better choices. But an individual can overcome a poor decision. For example, it may depart on a day that is less than ideal, but if it serendipitously finds a suitable lake with plenty of fish, it can overcome leaving on a poor day. However, cumulative poor decisions can prove fatal. Loons breeding near the West Coast fly west to the Pacific Ocean, and those breeding near the East Coast fly east to the Atlantic Ocean. Many loons farther inland, in Minnesota or Ontario, for example, rather than flying directly east or west, the shortest distance to the ocean, invariably fly southward to the Great Lakes. For loons in the interior, better decision making regarding the timing of departure and arrival is likely more critical to their survival and reproductive success than loons breeding nearer the coast. Loons banded in Minnesota and Wisconsin winter in both the northeastern Gulf of Mexico (in Texas, Alabama, and the panhandle of Florida) and the Atlantic Ocean (in both Carolinas, Georgia, and the Florida coast). I suspect this pattern may also be true for loons breeding in Manitoba. These data suggest that wintering in one location over the other confers no appreciable survival or fitness advantage, though this warrants further investigation. Which brings us back to the loon in the hotel bathtub.

In fall 2011, my good friend and colleague Darwin Long IV and I were examining the referenced locations produced by two loons implanted with satellite transmitters off the Louisiana coast. We discovered that they spent the summer in Saskatchewan. I remembered that the loons we caught at Walker Lake also went to Saskatchewan and was curious about the proximity of these two populations. Darwin was able to obtain the GPS coordinates of those loons from Mike Yates of Boise State. He plotted both data, and to our astonishment, one of the loons from Louisiana visited the same waterbody (Peter Pond Lake) as one of the loons from Walker Lake—an amazing coincidence! Keep in mind, we caught one loon at random out of a thousand or so on Walker Lake, and another at random wintering off the Louisiana coast. These data suggest that loons using Peter Pond Lake and neighboring lakes are wintering in the Pacific Ocean and the Gulf of Mexico. Some loons from this area migrate southwest, and others southeast. Since the Pacific Ocean is a considerably shorter distance from western Saskatchewan than the Gulf of Mexico (about 1,100 miles versus about 2,400 miles), one would surmise that loons would choose the Pacific Ocean to winter on since it would mean less energy expenditure and possibly increased fitness. Why fly an additional 1,300 miles if you do not have to?

Using knowledge of the distribution and history of glaciers may help us resolve this paradox. A glacial sheet covered North America at least twenty times during the past 5 million years. During glacial maximum, all birds, including loons, were forced south because the ice sheet covered the land. Loons breeding in western Canada migrated to the southwest, and loons in eastern Canada to the southeast. The two populations became separated (scientists have suggested that there may have been a third interior population). Eventually the glacial sheets receded during interglacials, or warming periods, and loons in each of the two subpopulations expanded and occupied their former northern range, but the middle of the continent would have been the last place to recolonize. Loons in the southeast region (Gulf of Mexico, Atlantic coast) eventually expanded their numbers northward, using lakes as they became available. Similarly, loons in the southwest (Gulf of California, Pacific coast) also expanded northward, occupying available lakes in Washington, Montana, British Columbia, and Alberta. Eventually these two subpopulations would expand their ranges farther and meet in Saskatchewan. But why does one loon breeding in Saskatchewan winter off the Pacific coast, and another one winter in the Gulf of Mexico? I suspect it is because they are tracing the respective pathways of each subpopulation; that is, this is what they have always done (a historical constraint). Will breeding loons in western Saskatchewan that winter in the Gulf of Mexico learn from other loons in the region that winter in the Pacific and migrate there instead? Will they learn a new migration pathway? More time and research may give us an answer.

Of the eleven adult loons that I have caught in western Saskatchewan, the males all varied in size. I suspect the larger males fly to the West Coast and winter in the Pacific Ocean, and the smaller males fly to the Gulf of Mexico. We do not know what portion of loon migration is under genetic control and how much is facilitated socially, but we do know that migration routes can change rapidly in some birds. In our Louisiana pilot migration study, the male loon departed and took a more direct route north to Saskatchewan. The female, however, flew to the East Coast and spent a week around Chesapeake Bay, before heading northwest to Saskatchewan. The male's migration distance was 40 percent shorter than the female's, and given the energy costs associated with flight duration, it would seem, once again, that the shorter distance would be the optimal strategy. Why would the female take the much longer route? To answer this question, we have to ask another: where do juvenile loons spend their summers?

The great majority of surviving loon chicks are not observed on or near their natal lake until their third or fourth summer. Many spend the summer on the ocean. In 2013 and 2014, the Biodiversity Research Institute conducted summer oceanic wildlife surveys using a digital camera mounted on an airplane. The plane flew in transects over a large part of the mid-Atlantic Ocean, and all wildlife was identified. I was curious to know whether they found any loons there during the peak breeding season (June 15 to August 15), because if they had, these would have to be subadults without territories. We counted sixty subadult loons, not one in breeding plumage. Some subadults may utilize large lakes within the breeding range with lots of neutral, or unclaimed, territories, lakes at the margin of the breeding range, or lakes outside of their range altogether (i.e., reservoirs, Great Lakes). If subadult loons were to return to their natal lakes, one of their parents or another adult would likely challenge them. Consequently, they would have less time for foraging but, moreover, could be hurt. Natural selection has likely shaped and prolonged the returning of subadults for a few years. The subadults have a safe place where they can pursue and catch prey without having to contend with überaggressive adults. Therefore, if that female from Saskatchewan spent its second (or even its third) summer in the mid-Atlantic, it retraced its steps to Chesapeake Bay because of its previous familiarity with it. The loon would have knowledge about the type and whereabouts of food, places protected from storms, and so on. I think the same reasoning applies to the male, but instead of migrating and summering in the ocean, it may have hooked up with other young loons, flown to the Great Lakes, and retraced its spring route to Louisiana, bypassing the ocean altogether.

Does sex play a role in the length of a migration? Kevin Kenow did not find any difference in migration distance between male and female loons breeding in the upper Midwest, but more data are needed for loons breeding closer to the coasts for conclusive results to be drawn. For example, though we have data from only five loons (two males, three females) in our Maine pilot study, males migrated shorter distances than females. The males wintered off the coast of Maine, a distance of 118 and 142 miles, whereas the females wintered south of Cape Cod (250 miles), off New Jersey (about 435 miles), and in Chincoteague Bay, Maryland (about 550 miles).

Spring and Fall Staging Areas

•

On both North American coastlines, the timing of spring loon migration is linked to the spring movement of marine fish. Several East Coast marine species, such as smelt, alewife, and river herring, blueback herring, and American shad, migrate upriver to spawn and lay eggs during February, March, and April, respectively. Historically, these spring runs of spawning fish were extraordinary in their abundance and have likely shaped the loon's spring migration. Loons gather at river mouths, often in groups, and often foraging cooperatively. Similarly, the Great Lakes and scattered reservoirs throughout the continent have large populations of landlocked alewife and herring that loons prey on during their annual spring migration trek.

The Great Lakes are an important staging area for loons in the upper Midwest and central Canada. They arrive in the northern part of Lake Michigan in late September and early October. Here, they are often observed as singles (not in groups), 5 to 6 miles offshore, in water about 100 feet deep. They remain there for roughly two weeks before flying to southern Lake Michigan, where they may stay an additional two weeks, often more than 10 miles from shore. Loons typically leave southern Lake Michigan in November (mean departure date is November 23). The average stay is twenty-seven days, but those that arrive before October 1 stay an average of sixty-eight days, and those that arrive after November 1 stay on average eight days. Here they find an abundance of food and fatten up before yet another leg of their migration journey, either to the Gulf of Mexico or to the Atlantic coast. However, in recent years these loons have been exposed to botulism.

Botulism in Loons

•

Scientists have known about avian botulism (type C) since the 1900s, and it was documented initially in waterfowl, such as mallards. But a different form (type E) was documented in fish-eating birds, such as mergansers, grebes, and loons, in the 1960s. Altogether there are seven different types of botulism (types A–G) based on characteristics of the neurotoxins produced by the bacterium *Clostridium botulinum*. Botulism neurotoxin may be one of the most toxic substances known. The bacterium reproduces by producing spores, which are released in the environment and settle in muck at the bottom of pools and lakes (the spores can also be found in the intestinal tracts of live animals). These spores can remain for years in sediment, resistant to extreme temperatures and severe desiccation, but under favorable conditions, such as an available nutrient source and proper pH and water temperature, they enter a vegetative state during which they grow and reproduce. The toxin is produced when bacteria grow during this vegetative state. Decaying animal carcasses and vegetation provide favorable conditions for *Clostridium* since the decay process uses up oxygen and creates anaerobic conditions. Botulism type E has caused periodic outbreaks of fish-eating bird mortality in the Great Lakes since the 1960s, but since the turn of the century, outbreaks have occurred more frequently.

These near-annual events of botulism were new and so was the devastation left in their wake. In a span of ten years, from 2000 until 2009, more than twenty thousand loons died from type E. The increase is due to changing weather patterns and non-native mussels and fish. Zebra and quagga mussels, both non-native species, became established in the Great Lakes in the late 1990s. These mussels, native to Eurasia, are common in the Black and Caspian Seas. They filter water more effectively than native mussels, allowing sunlight to penetrate farther into the water, prompting plant growth. Plants die and are decomposed by bacteria, which use up the oxygen, creating anoxic conditions that are ideally suited for the growth of *Clostridium* spores in the lake bed. As the mussels filter the water, they concentrate the toxin. This is where fish come into the story. The round goby is a small (4 to 9 inches), bottom-dwelling fish native to Eurasia. They feed on mussels but also crustaceans, small fish, worms, and eggs. Gobies were detected in the Great Lakes in the 1990s, having probably arrived via ballast from a foreign freighter. They were reported in Lake Erie in 1993 and became widespread throughout the lake by 2000. When a goby eats an infected mussel, it picks up the botulism toxin. Loons and other piscivorous birds feed on the infected gobies, because they are easy targets, and consequently die. The dramatic increase in the goby population happened to coincide with the beginning of more widespread botulism outbreaks in Lake Erie. Researchers have linked botulism in mussels with botulism in gobies and with botulism in loons; however, other fish have also tested positive for type E botulism, such as the freshwater drum, channel catfish, yellow perch, and lake sturgeon. So, while the round goby does play a role in current outbreaks, it appears not to be the only source of botulism. Furthermore, the frequency of type E botulism in the Great Lakes is related to warmer surface water temperatures and lower water levels. Thus, climate change has also likely contributed to the frequency of botulism outbreaks.

Loons migrate out of necessity, and selection has shaped their pattern and movements. They migrate to take advantage of environmental and biological conditions, and should those conditions change to the point where they no longer provide a benefit, loons will adapt, if suitable alternatives exist. Such was the case at Walker Lake. In 2001, I was the featured loon speaker at the annual loon festival in nearby Hawthorne, Nevada. I was excited to see so many people and families coming out to celebrate loons. However, the lake level had been dropping for forty years, and the amount of total dissolved solids in the water was increasing—sort of like a puddle drying up on a hot summer day. Farms and ranches located upstream had access rights to the water and could siphon it off to use as they deemed fit. Unfortunately, more water was siphoned annually than what came off the mountains in a normal year. This situation coupled with low precipitation and warmer than average annual temperatures caused evaporation to increase, and the lake level dropped approximately 181 feet between 1882 and 2016. The concentration of solutes in the water was becoming alarmingly high. A local group of citizens concerned about the status of the fishery and health of the lake started the Walker Lake Working Group. Despite their efforts, lake levels continued dropping, and the fish could no longer pump the thick, sludgy water through their gills to get the requisite oxygen for survival. They quickly died. As of 2016, the total dissolved solids concentration reached 26 grams per liter, well above the lethal limit for most of the native fish species throughout much of the lake. Lahontan cutthroat trout are no longer present in the lake, and tui chub have declined dramatically and may soon also disappear. With the decline of the fishery, the loons left. The last loon festival in Hawthorne was in 2008. The case of Walker Lake is yet another example of how climate change is drastically altering patterns of life.

Further Reading

Alerstam, T., A. Hedenstrom, and S. Akesson. 2003. Long-distance migration: Evolution and determinants. *Oikos* 103: 247–60.

Alerstam, T., M. Rosén, J. Bäckman, P. G. P. Ericson, and O. Hellgren. 2007. Flight speeds among bird species: Allometric and phylogenetic effects. *PLOS Biology* 5 (8): e197. DOI: 10.1371/journal.pbio.0050197.

Boag, P. T., and P. R. Grant. 1981. Intense natural selection in a population of Darwin's finches (*Geospizinae*) in the Galapagos. *Science* 214: 82–85.

Chipault, J. G., C. L. White, D. S. Blehert, S. K. Jennings, and S. M. Strom. 2015. Avian botulism type E in waterbirds of Lake Michigan, 2010–13. *Journal of Great Lakes Research* 41: 659–64.

Cramp, S., and K. E. L. Simmons. 1977. *The Birds of the Western Palearctic.* Oxford: Oxford University Press.

Essian, D. A., J. G. Chipault, B. M. Lafrancois, and J. B. K. Leonard. 2016. Gut content analysis of Lake Michigan waterbirds in years with avian botulism type E mortality 2010–2012. *Journal of Great Lakes Research* 42: 1118–28.

Gostomski, T. J., and D. C. Evers. 1998. Time-activity budget for Common Loons, *Gavia immer*, nesting on Lake Superior. *Canadian Field-Naturalist* 112: 191–97.

Gray, C., J. D. Paruk, C. R. DeSorbo, L. J. Savoy, D. E. Yates, M. Chickering, R. B. Gray, et al. 2014. Strong link between body mass and migration distance for Common Loons (*Gavia immer*). *Waterbirds* 37: 64–75.

Hedenström, A., and T. Alerstam. 1997. Optimum fuel loads in migratory birds: Distinguishing between time and energy minimization. *Journal of Theoretical Biology* 189: 227–34.

Hedenström, A., and T. Alerstam. 1998. How fast can birds migrate? *Journal of Avian Biology* 29: 424–32.

Johnson, D. 1997. Wing loading in 15 species of North American owls. In *Biology and Conservation of Owls of the Northern Hemisphere: 2nd International Symposium,* Gen. Tech. Rep. NC-190, 553–61. Edited by J. R. Duncan,

D. H. Johnson, and T. H. Nicholls. St Paul, Minn.: U.S. Department of Agriculture, Forest Service.

Kenow, K. P., P. D. Adams, N. Schoch, D. C. Evers, W. Hanson, D. Yates, L. Savoy, et al. 2009. Migrations patterns and wintering range of Common Loons breeding in the northeastern United States. *Waterbirds* 32: 234–47.

Kenow, K. P., S. C. Houdek, L. J. Fara, B. R. Gray, B. R Lubinski, D. J. Heard, M. W. Meyer, T. J. Fox, and R. J. Kratt. 2018. Distribution and foraging patterns of Common Loons on Lake Michigan with implications for exposure to type E avian botulism. *Journal of Great Lakes Research* 44: 497–513.

Kenow, K. P., M. W. Meyer, D. C. Evers, D. C. Douglas, and J. Hines. 2002. Use of satellite telemetry to identify Common Loon migrations routes, staging areas and wintering range. *Waterbirds* 25: 449–58.

Kerlinger, P. 1982. The migration of Common Loons through eastern New York. *Condor* 84: 97–100.

Lafrancois, B. M., S. C. Riley, D. S. Blehert, and A. E. Ballman. 2011. Links between type E botulism outbreaks, lake levels, and surface water temperatures in Lake Michigan, 1963–2008. *Journal of Great Lakes Research.* 37: 86–91.

McIntyre, J. W. 1988. *The Common Loon: Spirit of Northern Lakes.* Minneapolis: University of Minnesota Press.

Newman, S. H., A. Chmura, K. Converse, A. M. Kilpatrick, N. Patel. E. Lammers, and P. Daszak. 2007. Aquatic bird disease and mortality as an indicator of changing ecosystem health. *Marine Ecology Progress Series* 352: 299–309.

Paruk, J. D., D. C. Evers, J. W. McIntyre, J. F. Barr, J. N. Mager, and W. H. Piper. 2021. Common Loon (*Gavia immer*), version 2.0. In *The Birds of North America,* edited by P. G. Rodewald. Ithaca, N.Y.: Cornell Lab of Ornithology. https://doi.org/10.2173/bna.

Paruk, J. D., D. Long IV, S. L. Ford, and D. C. Evers. 2014. Common Loons wintering off Louisiana Coast tracked to Saskatchewan during the breeding season. *Waterbirds* 37: 47–52.

Pennycuick, C. J. 1975. Mechanics of flight. In vol. 5 of *Avian Biology,* edited by D. C. Farner, J. R. King, and K. C. Parks, 1–75. New York: Academic Press.

Poole, E. L. 1938. Weights and wing areas in North American birds. *The Auk* 55: 511–17.

Riley, S. C., K. R. Munkittrick, A. N. Evans, and C. C. Krueger. 2008. Understanding the ecology of disease in Great Lakes fish populations. *Aquatic Ecosystem Health Management* 11: 321–34.

Waldman, J. 2013. *Running Silver: Restoring Atlantic Rivers and Their Great Fish Migrations.* Guilford, Conn.: Lyons Press.

10

Not Your Typical Snowbird

The Loon's Winter Ecology

In the winter of 2014, I was leading a group of four Earthwatch volunteers on a night outing to catch Common Loons in the Gulf of Mexico. We were there because I was investigating the potential effects the Deepwater Horizon oil spill may have had on loons in the area. Were they exposed to the oil? And if so, at what concentrations? Did those concentrations affect loon health, and if so, how? The objective was simple: catch loons, collect tissue samples, and analyze them for oil. My lead field assistant, Hannah Uher-Koch, was the primary loon spotter and the one running the day-to-day operations. I had just flown down for a few days, but she had been there all winter long and knew better than I where to catch loons. She was an outstanding field assistant, extraordinarily competent, organized, and enthusiastic. Our driver and boat captain was Todd Seither, a local fishing guide. Todd was a great asset to our team because he, too, was skilled at spotting loons, but moreover he was conscientious, always looking out for our safety. The pressure was on me, as the netter, to get the loons in the net (easier said than done).

We loaded the gear into Todd's boat, a sleek 24-foot skiff, but in doing so, we alarmed a local Black-crowned Night Heron. As it flew overhead, it let out a distinctive, loud barking call. With our gear in tow, we were slightly cramped in the boat as we assumed our positions, Hannah and me up front, followed by the volunteers. It was pitch dark at 7:45 p.m. The night was mild, the temperature about 50°F, with low wind and good visibility. We had been out for twenty-five minutes when Todd noticed a buoy in the water moving against the tide and said to me, "There must be some fish attached to it." He told the group that some anglers bait a large hook and attach it to a buoy in the hope of landing an alligator gar, a large fish with many teeth. He pulled the boat next to the bobbing buoy and instructed me to pick it up. To get leverage, I laid myself flat across the bow of the boat and leaned my arms and head over the sides to lift the buoy. It was dark, and I could not see. I lifted. It was not an alligator gar—it was worse, much worse. Adrenaline shot through my body. I remember it well to this day, for out of the water lunging at me was a 5-foot-long sand shark. At that moment I asked myself, "Now, why am I out here?" We had come to study the effects of a major environmental event on wintering loon populations and had a shark dangling off our boat. I clearly had a lot more to learn.

Marine Habitat Selection

•

If we plot data on the winter distribution of Common Loons published by the National Audubon Society and the U.S. Fish and Wildlife Service, we can easily detect a clear pattern. In the winter, the Common Loon is observed up and down the entire East and West Coasts of North America, the northern coast of the Gulf of Mexico as far south as the Yucatan Peninsula, and within the Gulf of California. The loon is the national bird of Canada, and we associate it with boreal forests and spruce trees, yet we can also see it wintering off the coast of Mexico, among yuccas, cacti, and palms. I observed this odd juxtaposition firsthand twelve years ago. I was an associate professor of biology at Northland College in Wisconsin teaching a sixteen-day field ornithology course in Mexico. I designed the course so we would spend some time along the Baja Peninsula (why not, right?). While there, I spied a loon swimming parallel to the shore and excitedly pointed it out to the class. It remains one of my more memorable sightings, a loon swimming calmly on the surface with tropical marine birds, frigatebirds, and boobies flying overhead, and cacti and other desert succulents in the backdrop.

Coastlines of North America vary in their shape and size. Bays and coves are common, as are landmasses that protrude, such as Cape Cod. Rivers empty into the ocean. Some, such as the Mississippi, are large and silty; and others, such as the Androscoggin in Maine, are small but relatively clear. Estuarine water, where fresh and marine waters mix, is generally supportive of abundant plant and animal life. Some coastal waters are relatively shallow, less than 100 to 200 feet for 100 or more miles offshore, but in some cases, the continental shelf does not extend as far, and a deep trench, bringing cold and nourishing water, occurs closer to shore. All of these factors are important to loons. In our quest to understand what ecological factors impact loons, we will compare a few locales. Two important wintering areas for loons on the West Coast are Puget Sound in Washington and Monterey Bay in California, and two on the East Coast are Cape Cod, Massachusetts, and Pamlico Sound in North Carolina.

In 1792, Peter Puget, a Huguenot lieutenant, accompanied George Vancouver on an expedition to the Pacific Northwest. To honor him, Vancouver named the majestic sound in northwestern Washington after him. Puget Sound is a large body of protected water, 450 feet deep, with both marine and freshwater (nine major rivers empty into it) and surrounded by the Olympic and Cascade Mountains. It is a fjord system of flooded valleys that were covered by glaciers numerous times in the past 5 million years. Puget Sound also marks the beginning of the inlet passage to Alaska. This area experiences big tides, with the difference between high water and low water around 8.3 feet. Many marine mammals are found there, including orcas, seals, and otters. The many fish species include the recognizable salmon, trout, and char and the less familiar Pacific herring, surf smelt, and Pacific sand lance. Because of this abundance, seabirds such as Pigeon Guillemots, Rhinoceros Auklets, Common Murres, and Marbled Murrelets abound. In addition, other waterbirds such as Surf Scoters, grebes, and cormorants, and non-waterbirds, such as Bald Eagles, are in high abundance. Loons, over a thousand of them, are abundant there too. We can deduce that Puget Sound is a preferred wintering location for loons for two reasons, one biological and the other physical. First, loons travel to where the food is, and the Puget Sound supports a tremendous fishery. Second, the physical appearance and shape of the sound protects and buffers loons and other wildlife from large surging storms and high winds—it is a safe haven. Given the choice of resting in a bay or sound or out in the open ocean, most loons on most days seem to opt for the former.

California's Monterey Bay is not nearly as protected from the open ocean as Puget Sound, but about 10 miles offshore is Monterey Canyon, the largest submarine canyon along the West Coast of North America. At 12,173 feet, it is deeper than the Grand Canyon, and due to an upwelling of cold, rich nutrients, it provides a base for supporting an abundance of plankton, which in turn feeds numerous fish populations and an incredible variety of marine life. Probably most famous are the gray and humpback whales, which stop by Monterey Bay biannually and feed during their migration. Bottlenose dolphins, harbor seals, and sea otters are also abundant. This area supports about 500 different fish species and more than 180 species of seabirds and shorebirds, including loons. Show me large schools of baitfish, and I will show you a loon! Monterey Bay is a unique place for loon watchers, and observing loons and dolphins together in the same field of view is another of those odd life juxtapositions that enrich a loon biologist's world. Now on to Cape Cod and Pamlico Sound on the East Coast.

Cape Cod is a peninsula that jets 65 miles out from the Massachusetts coastline, ending in an L shape, with the northern extension of land curling around and enclosing Cape Cod Bay. It shields much of the Massachusetts coastline from North Atlantic storm waves. Cape Cod Bay receives lots of cold water from the Gulf of Maine and is home to seals, dolphins, and whales. Historical accounts describe abundant whale populations in Cape Cod Bay. Commercial and recreational fishing is common, and there is an abundance of bird life, including loons. Although Cape Cod Bay is not as protected from the open ocean as Puget Sound, it offers more protection than Monterey Bay and provides a suitable wintering habitat for loons. South of the Cape Cod peninsula is Nantucket Sound, a body of water enclosed by a couple of large islands. Nantucket Sound is located at the confluence of the cold Gulf of Maine and the warm Gulf Stream and, as such, is rich in nutrients and home to many marine mammals and seabirds.

The central part of North Carolina bubbles out and ends in a series of estuaries (80 miles long and 15 to 20 miles wide) called Pamlico Sound. It is the second largest estuary in the United States—only Chesapeake Bay is larger. Pamlico Sound is separated from the Atlantic Ocean by the Outer Banks, a row of low, sandy barrier islands, including Cape Hatteras National Seashore. The sound supports commercial fishing and a large shellfish industry. It is home to many waterfowl and seabirds, especially in the winter. Due to its protection from surging storms and waves, and its abundance of food, Pamlico Sound also supports a large wintering population of Common and Red-throated Loons.

The number one characteristic that unites these four prominent wintering sites for loons is a rich abundance of food, primarily fish, coupled with water clarity. Without suitable prey and clear water, loon populations would be appreciably smaller or absent altogether. Loons may vacate the mouths of coastal rivers if these waters become too turbid, reinforcing the importance of water clarity to wintering loons. Loons do occupy estuarine turbid waters, as they do in Louisiana, for example, but they are not as abundant. When gathering some much-needed data on biological and environmental variables that might predict the winter distribution and abundance of loons, researchers from Rhode Island found that the surface water concentrations of chlorophyll *a* were a good predictor of loon presence and abundance. Chlorophyll *a* is used by green plants (plankton) to photosynthesize and is a proxy for biological activity. In other words, where there is phytoplankton, there are zooplankton and fish. And where there are fish, piscivores such as loons are soon to follow.

The second important characteristic common to the four predominant loon wintering sites appears to be shelter that offers protection from surges associated with open water. Loons are very comfortable in the open ocean. I have observed them 12 miles from the mainland, and other researchers have reported them 50 to 60 miles from shore. The distance from the shoreline to where loons are found is likely associated with the proximity of the ocean shelf to the coastline. In the Gulf of Mexico, the shelf is more than 100 to 150 miles from shore, and loons in this area are commonly found 60 miles from shore. But in northern New England and Monterey Bay, where the ocean shelf is nearer (10 to 50 miles), loons are not found as far out as they are in the Gulf of Mexico. In southern New England, loons are more prevalent closer to shore in water less than 110 feet deep.

What is a loon doing 50 (or more) miles from shore? Feeding. The water depth over the continental shelf varies (50 to 500 feet) but often is between 200 and 300 feet—deep, to be sure, but not so deep to prevent a loon from reaching the bottom to find food. If given the choice, do loons prefer the open ocean miles from shore or being closer to the coast? The data on this are incomplete, mostly because few surveys of loons miles from shore in the middle of winter have been done. Although loons far offshore may go undetected and be undercounted compared to those adjacent to the coast, I suspect that loons are better off nearer shore. Here's my reasoning. Ecological theory predicts that loons will inhabit areas where they can maximize their caloric intake while minimizing their caloric expenditure. If a loon in the open ocean has to dive 200 feet to find food and another loon closer to the coast can do the same in 40 feet of water, the latter strategy is more likely to be selected. Granted, more loons mean more competition for food, but based on spacing patterns and behaviors I have observed along coastlines, I suspect it is not intense enough to drive loons to the open ocean. However, when waters near the mouths of rivers become turbid from storms and runoff, loons may have more difficulty locating prey. Under these conditions, loons may do better farther away from shore. But winter storms are common occurrences in some locations, and in those circumstances a loon may be better off in a bay or cove, nearer the coastline and protected from the large waves, than in the open water.

If loons, like pigeons, can detect air pressure changes, then they may be able to move before a major storm hits, but they cannot fly during winter. Loons undergo a winter wing molt that renders them flightless. In all my work, I have never seen an adult Common Loon in flight during the months of December through February. A loon must paddle along the surface to move from one location to the next. Moving at roughly 3 miles per hour, a loon could theoretically swim 10 to 15 miles to shore in three to four hours, should the urgency require such effort.

Depending on seasonal storm patterns, some loons may move closer to shore rather than farther away from it for two reasons. First, remember that most shorelines are irregularly shaped, with bays and coves that provide refuge for loons compared to the open ocean. During a storm, a loon would spend considerably more energy maintaining its position in the water column in open water than in a protected bay or cove. In winter 2016, a team of biologists working for me in the Gulf of Mexico observed the effects of a huge windstorm firsthand on nearshore loon abundance. Prior to the storm, Todd Seither was out doing surveys and saw only the occasional loon, but the next night after the storm event, loons were everywhere! This observation suggests that loons move to more protected coastal waterways in response to surging storms and winds to save energy. Second, forage fish in the size class preferred by loons may be more abundant in bays or coves than in the open ocean. Both Puget and Pamlico Sounds are excellent examples of areas well protected from storms that also support these fish populations.

In summary, there are two biological variables (fish abundance and chlorophyll *a*) and two environmental variables (protected shorelines and water clarity) that appear to be good predictors of winter loon distribution and abundance. Yet, there is still more to investigate. For example, how does the level of tidal fluctuation influence the distribution or abundance of loons wintering along the coasts, and why do some loons spend their winter on freshwater reservoirs instead of marine coastlines?

Freshwater Habitat Selection

•

Roughly 5 percent of North American Common Loons overwinter on reservoirs instead of going to the ocean. Across North America, especially in the Southwest and Southeast, thousands of loons use dozens of reservoirs as a final winter destination spot. They include, for example, Lake Mead, Lake Powell, Lake Hartwell, Lake Jocassee, and Lake Strom Thurmond. In addition, tens of thousands of loons use reservoirs as stopover locations during migration. Loons are likely wintering at these reservoirs for the same reasons they choose certain marine habitats: abundance of prey, water clarity, and protection from storm events. For the past six winters, I have been going to Lake Jocassee reservoir in South Carolina to study the winter ecology of the Common Loon. Approximately 150 Common Loons return to the reservoir each winter, and many more visit during spring migration. I wanted to investigate the distribution of loons here because it seemed nonrandom. More were found on the northeast end, and considerably fewer on the northwest end. Teaming up with Jay Mager from Ohio Northern University, Brooks Wade of Jocassee Lake Tours, and a team of Earthwatch volunteers, and by sharing data with Duke Energy fishery biologists, we began to get a clearer understanding of loon distribution on the reservoir.

First, we confirmed that the distribution of loons on the reservoir was indeed nonrandom. Second, we correlated loon density with forage fish density. Lake Jocassee has a large forage fish population of blueback herring and threadfin shad, ideal for loons because of their narrow shape and relatively small size. Third, we found that loons were also more prevalent in areas with river mouths than without. Forage fish like herring and shad spawn in rivers and creeks, and the survival of their young depends on a rich plankton community, which is largely regulated by nutrient loads carried by rivers. We often observed large numbers of loons flock feeding at the river mouths, with one exception, and this puzzled us. Learning the history of the lake helped us understand this observation. The watershed draining into Jocassee's northeast arm was cut over for agriculture and dwellings, whereas the watershed above the northwest arm remains pristine and undeveloped. We suspect that there is a pronounced difference in the amount of nutrient runoff between the two watersheds, and this likely accounts for the aggregation of loons near the mouths of rivers in the northeast zone. This zone had significantly higher chlorophyll *a* concentrations than any other area on the lake. Thus, the difference in the number of loons we observed between the two arms of the reservoir is likely accounted for by the difference in the amount of human disturbance (cutting of trees) in one area compared to the other, a sad realization. Finally, the water at Lake Jocassee is exceptionally clear (more than 26 feet), and loon foraging success increases with water clarity. Based on this, it appears that loons select reservoirs for the same reason they select marine habitats.

Winter Site Fidelity and Winter Movements

•

In 1997, I met Darwin Long IV at a loon conference in St. Paul, Minnesota. I knew of Darwin and his work, knew that he was from Southern California and loved loons, and I was looking forward to exchanging ideas with him. We are fortunate if in our lifetimes we meet someone with a similar passion for a subject, hobby, or interest. Darwin could talk about loons and their biology all day long (and I thought I was bad!). He wanted to study wintering loons in Morro Bay, California, so we discussed graduate school and outlined a project. We kept in touch, and soon I was taking a group of students from Feather River College with me to Morro Bay in 2001 to help Darwin set up his research. The project was a success, and over the next six years, Darwin caught and uniquely color-banded eighty loons. Throughout his winter surveys, he was able to track individuals and ask some important questions: How long do they stay? Do the same individuals come back each year? In other words, do they exhibit winter site fidelity? Of the twenty color-banded adults, fifteen were observed again in subsequent winters, and twelve were observed over multiple winter seasons. One loon banded in 2004 was observed in Morro Bay for six consecutive winters. This same bird was observed again in 2020, indicating that it continued to remain loyal to its 2004 winter habitat sixteen years later. These data were the first to show that loons exhibit winter site fidelity.

In a twist of fate, Darwin and I teamed up ten years later, in 2011, to study loons wintering off the Louisiana coast. Dave Evers at Biodiversity Research Institute (BRI) asked us both to help with the Deepwater Horizon oil spill environmental assessment on wintering waterbirds. I developed a long-term research project investigating the exposure rate and the effects of exposure to oil on Common Loons wintering in the area. Over seven consecutive winters my team caught 128 loons, plus 10 recaptures. Several of the recaptured loons were in or near the original location where they were banded a year earlier, suggesting that they, too, exhibit winter site fidelity, just like the loons in Morro Bay. Using Darwin's long-term data on winter site fidelity from California, and mine from Louisiana, Evan Adams, a quantitative ecologist at BRI, developed a model and crunched the numbers. The results confirmed what we had suspected: Common loons exhibit winter site fidelity. Approximately 85 percent of adults return to the same winter location on a consistently regular basis.

In 2011, I was involved in another BRI project to study the migration and winter movements of loons breeding in Maine (see chapter 9). We implanted six satellite transmitters, and although one failed, two lasted for more than eighteen months. These longer-lasting transmitters afforded us a unique opportunity to compare each individual's winter movements between years. Both loons migrated to the same area of the ocean in consecutive years. Moreover, one individual had a 100 percent overlap in its winter use area from one year to the next, and the other loon's overlap was 64 percent. Collectively, these data support the notion that if you walk along a particular stretch of beach in the winter and see a loon, it is likely the same bird you saw the previous year.

Why would a loon go back to the same location each year? Probably for the same reason many of us vacation at the same campground or cabin each year. We return to these same locations, in part, because we are familiar with them and comfortable there. We know the location of the showers, the great vistas, the trailheads, the best fishing spots, and the place that serves delicious ice cream. Because of that familiarity, we have fewer decisions to make, and we have more time to relax. For a wild animal, that time gained translates into more time for foraging, important if it is to survive the winter in good condition. Winter site fidelity developed in loons, in part, because individuals that returned to the same area annually gained valuable local knowledge about their environment, such as the best locations to find food, to seek refuge from a storm, and to evade predators. Over time, these choices became fine-tuned and improved a loon's body condition and survival rate.

Like adults, immature loons exhibit winter site fidelity but perhaps not to the same extent. If this supposition turns out to be true, then identifying what factors are responsible for them switching winter locations between years may be important for conservation managers to know, mostly because immature loons experience higher mortality in the winter compared to adults.

Daily Activity Patterns

•

There is less daylight in winter compared to summer, and wintering loons have to get on with the business of finding food. When I have watched loons, either from shore using spotting scopes or from a boat with binoculars, I see them typically awake at first light, stretch, swim to a preferred foraging location, and commence diving. They feed throughout the day, with distinct feeding bouts lasting between forty-five and sixty minutes and consisting of either short- or long-duration dives. After each foraging bout, they rest on the surface and often preen. Preening is a common activity, even more so than in the summer. This cycle of eating, resting, and preening continues throughout the day. A few individuals may paddle on the surface to a different location and dive there. We followed one such loon swimming in the Gulf of Mexico and tracked it using GPS. It paddled 2.2 miles, before diving and feeding in a completely different area. In early afternoon a bird may close one, or both, of its eyes for a few minutes, but come late afternoon or early evening, loons often begin to aggregate and swim out to deeper water.

Around dusk, a group will sometimes form a raft, essentially a group of loons that associate together at night. Rafts vary in size, from small groups of four to twelve individuals to larger ones of twenty to forty (sometimes fifty to one hundred) individuals.

The fact that loons often raft at night suggests there is some benefit to this behavior. How it benefits them is a difficult question to answer because it is hard to establish a control and experimental group. But we can use inference based on our observations. One explanation is that being in a raft is statistically safer than remaining alone. In marine environments, large underwater predators, such as sharks, have been known to kill loons. One loon chick Dave Evers banded at Seney National Wildlife Refuge washed up on the shore of South Carolina in midwinter with razor-like cuts and a shark tooth embedded in its body. Furthermore, being away from shore minimizes the chances of being left stranded in receding tides or being washed ashore during storms. So, it is possible that loons are safer rafting farther offshore and in deeper water. Collectively, some or all these factors may play a role in rafting behavior in loons. The motivation to raft must be deeply ingrained in loons because I have observed them raft on reservoirs that are without tides, currents, or sharks. This suggests to me there are some benefits to loons rafting, such as higher survivorship, and that natural selection has shaped this behavior over time.

Loons that inhabit areas with greater tidal fluctuation are more likely to have their circadian rhythms influenced by the tides than those living in regions with little tidal influence. For example, loons wintering off the coast of Maine (about 45° latitude) experience daily tidal fluctuations of 10 to 12 feet, but those wintering off Morro Bay, California, and Barataria Bay, Louisiana (about 25° latitude) experience a tide of roughly 1 foot. This striking contrast in tidal fluctuations influences a loon's movements. During receding tides, Maine loons move closer to shore, where prey are more concentrated and potentially slowed in their evasive movements. In addition, the fish that loons are targeting may be moving inshore in response to the more exposed food. Such pronounced movements of fish do not occur in Barataria Bay, for example. Igor Malenko, a graduate student at the University of Southern Maine, suspects that groups of loons develop a dominance hierarchy during the winter. If this is true, it opens up a new set of questions to investigate.

Time-Activity Budgets

•

If you wanted to compare behaviors between breeding and wintering loons, or between loon populations from two different habitat types, how would you go about it? Niko Tinbergen, a Nobel Prize–winning Dutch scientist, understood the value of recording animal behavior in such a way that comparisons can be made between individuals, sexes, or populations. These comparisons may reveal a pattern that might otherwise go unnoticed and, along the way, help us understand how animals use and partition their time. In the 1950s, Tinbergen developed and refined the time-activity budget (TAB), used by researchers to quantify behavior. Here's how it works: Break the behaviors you are interested in recording into discrete categories; for loons, this would be foraging, preening, resting, and locomoting (i.e., swimming on the surface from point A to point B). Next, record behavioral events that may be of significance, such as vocalization, a foot waggle, and a wing flap. Then adopt a methodology. For example, you can record behavior continuously or at discrete time intervals, such as every one, two, or five minutes. I prefer the latter because the data are easier to analyze in this format, but each method has its pros and cons. Time-activity budgets can be used to investigate how animals vary their behavior over the course of a year (summer versus winter) or season (breeding versus nonbreeding) and to help identify what constraints animals face.

Summer versus Winter

To compare TABs of summer and winter loons, we need data on breeding loons without chicks. Winter loons do not have chicks; thus, observations of summer loons that are tending chicks would require recording an additional category (chick feeding), invalidating the comparison. These data exist, and investigators report that loons during early spring and midsummer (without chicks or young) foraged and preened roughly 50 to 54 percent and 10 percent of the time, respectively. Wintering loons spent more time both foraging (54 to 60 percent) and preening (14 to 24 percent) than summer adults. I think winter loons spend a greater percentage of their daylight hours foraging because they have less time in which to do so (approximately 11 versus 14.5 hours, but this will vary with latitude). But they may also spend more time foraging because they are molting and replacing their feathers, which requires additional energy. This new feather growth may also be the reason that loons in the winter spend a greater proportion of their daily time preening compared with loons in the summer. Because new feathers must be aligned and manipulated to create an impenetrable water barrier, preening is an important activity for winter loons.

Unlike waterfowl, which undergo a wing molt prior to migration, loons delay their body and wing molt until they reach their winter destination. Rather than molt or replace their wing feathers gradually, they lose all their flight feathers simultaneously, becoming flightless for a period of two or three weeks. A rapid wing molt over a few weeks is more stressful to individuals than a protracted molt over a longer period. However, it is likely advantageous to be flightless for a short period of time rather than enduring labor-intensive flight over a longer period. There is a lot of variation in the timing of the wing molt in both adult and immature loons in a given area. For example, one adult loon off the Louisiana coast was in the middle of its wing molt on February 9, while another adult was beginning its wing molt on March 5, nearly four weeks later. In Morro Bay, we found one adult molting its wing feathers on January 27, and another adult nearly a month later, on February 22. Why the difference in molting times?

Molting in an individual can be delayed by physiological stress and poor health, which best accounts for the differences in the molt timing of loons wintering at Morro Bay and Barataria Bay. Over a broad geographic range, the timing of the molt is influenced by latitude. Loons wintering farther south initiate their wing and body molt sooner than individuals wintering farther north. This is an excellent example of selection at work. Mid-north Atlantic waters are considerably colder than those in the Gulf of Mexico during the winter months, probably 25°F to 30°F cooler on average. Loons wintering farther north need to conserve energy, and since molting is energetically expensive, it is better to delay it until the water and daytime ambient temperatures warm up, such as in early spring. I found that loons wintering farther south (South Carolina) undergo their molt four to six weeks sooner on average than loons wintering farther north (Maine).

Loons off the Louisiana coast, near the mouth of the Mississippi River, spent more time foraging (60 to 65 percent) than marine loons off the coasts of Maine, Virginia, and Rhode Island and at a freshwater reservoir in South Carolina. This difference is likely due to the high turbidity in the Louisiana study area, which affected their foraging ability. Remember, loons are visual predators, and without clear water they likely experience challenges to meeting their daily caloric intake. When I placed my hand in the murky waters off the Louisiana coast, it disappeared completely 18 inches below the surface. I suspect that loons in these highly turbid, but shallow waters dive to the bottom and use their bills as plows, making their way through the substrate. Several times during my research I observed a loon surface with a flounder, a fish buried on the bottom, which can only be dislodged by probing.

What Do Loons Eat in Winter?

No detailed diet studies of what loons eat have been undertaken on their winter range, mostly because of the challenges involved. First, you must catch a loon (a tricky feat in itself) and then perform a gastric lavage, which empties the stomach contents. The technique consists of sticking a tube down the throat into the stomach, pumping a saline solution into the stomach, inverting the bird, and having it vomit into a pail. The few times I have assisted with this procedure we did not recover much, mostly because the loon stomach is so acidic that it quickly dissolves fish bones. A more recently developed technique consists of collecting fecal samples and analyzing the DNA in them to identify prey items. I have just started using this method in my research.

Loons are opportunistic. They eat a variety of fish species, of different sizes and shapes (e.g., haddock, anchovy, and flounder). They eat both schooling (e.g., menhaden and silversides) and nonschooling (Atlantic croaker and spot) types, as well as species that reside nearer the surface (e.g., alewives and herring) and those that prefer being nearer the bottom (gobies and flounder). Wintering loons also feed on invertebrates, predominately crab, but will eat shrimp on occasion. I have seen loons consume crabs readily while wintering in Southern California, Maine, and Louisiana. Many loons in the Gulf of Mexico, especially those near shore or in shallower water, will eat crabs all day long. I wonder if crab may not be a preferred food for some local loon populations.

Compared to a fish, a crab may be easier to locate and catch but requires a fair amount of handling time to delimb or declaw before swallowing. I once saw a crab's pincher lock on to the side of a loon's mouth, fighting for its survival, only to have the loon fling the crab off with great force into the air. The loon then swam back to the crab, severed off one pincher, then the other, before swallowing it. Judy McIntyre may have been the first to suspect that emaciated or young loons feed predominately on crabs because they are easier to locate and catch than fish. My observational data support Judy's idea; young loons do feed on crabs more often than adults in some local populations. However, in some areas, such as off the Maine coast, I have observed presumably healthy adult loons target crabs frequently. In fact, under certain environmental conditions, such as in the turbid estuarine waters off the Louisiana coast, a loon may find it more profitable to plow in the substrate than swim aimlessly in the hope of visually locating a fish. This strategy of foraging may be optimal if crabs are abundant, which they are in coastal Louisiana and Maine.

Foraging Behaviors: Solitary versus Social

In marine environments Common Loons do not defend a territory or typically forage as solitary individuals. Generally, birds explore an area and move on, sometimes coming back hours later to an area they searched previously. Young birds likely follow in the wake of adult birds, moving from one patch to the next. Dominance hierarchies exist in birds, and almost always the adult is dominant to the immature, as is likely the case in loons. Igor Malenko observed one loon spend forty-six minutes handling and trying to swallow a large fish it had brought to the surface (likely a sea bass), and Earthwatch volunteers have seen loons attempt to swallow a large fish for twenty-five to thirty-five minutes. In a lot of these cases, the loon has no return on its investment. The advantage of group foraging over solitary foraging is that individuals may work together to herd schooling fish, thereby increasing their odds of success. The more individuals foraging collectively, the more chaos ensues for their prey, and while the fish are busy eluding one loon, they fail to see the next one. The effectiveness of group foraging ultimately depends on the behavior of the prey, and whether they school or not (some fish are solitary). Small fish are swallowed whole and take little handling time, but if they are too small, loons need to catch a lot of them, which requires more time and effort. Cost-benefit analysis predicts a point of diminishing returns for a loon trying to fill its belly on small prey, unless they are superabundant and relatively easy to catch. Therefore, schooling fish, such as herring, alewives, and menhaden, are main components of a wintering loon's diet.

It is not uncommon to observe a large number of loons at river mouths in the middle of winter along the coast of Maine. On the afternoon of February 25, 2011, Darwin Long IV encountered an enormous flock of loons roughly 10 miles off the Mississippi shoreline in the Gulf of Mexico. Together we looked at his photographs and counted several hundred loons, though some were undoubtedly underwater. We concluded there were more than six hundred loons in the group, the largest loon foraging flock ever reported to my knowledge. This observation suggests that the formation of extremely large rafts of loons is opportunistic and based on the type of prey available. Other species of birds, such as Northern Gannets, Brown Pelicans, and various tern species, will sometimes join a loon foraging flock in the ocean. When Darwin checked on the flock of loons the next day, he found only a few solitary individuals remained; the flock had dispersed. As the food goes, so go the loons.

The Loons of Lake Jocassee

•

The Common Loons wintering at the freshwater reservoir of Lake Jocassee are unique in the high prevalence of social behavior they exhibit compared to marine birds. Jocassee loons tend to remain near and interact with other loons on a consistent basis. They stay in their respective basins throughout the winter; we have never observed banded or radioed loons from the northern part of the lake in the southern part. Moreover, the consistency of the number of loons in distinct basins documented in our weekly surveys suggests that these are the same individuals and that they know each other. This finding leads to several questions: How do these basin assemblages assort? Is it by contingency, or are other forces at play, for example, sex, relatedness, or being in a distinct subpopulation? Can individuals recognize each other as coming from different subpopulations? Do cooperative foraging groups fare better than individuals that remain solitary, and why are some individuals solitary and not in a group?

From our repeated winter surveys on Lake Jocassee, many of the loons remain in groups throughout the day, collectively resting, preening, and foraging together. This high degree of sociality is less common in loons wintering in the ocean (though at river mouths, marine loons can be social). I suspect the reason loons at Lake Jocassee socialize more than loons in marine environments has to do with the distribution of prey, in most cases, baitfish (or forage fish). In the ocean, tidal currents and wave action likely disperse zooplankton, which in turn changes the distribution of forage fish that prey on them, but in freshwater environments, tidal effects are absent, and zooplankton are less likely to be dispersed. Thus, forage fish may be more consistently located in reservoirs than in ocean or estuarine habitats, and this consistency and predictability allow loons to work together more frequently.

At Lake Jocassee, when baitfish, like shad, are nearby or at the surface, they form bait balls, which are easily observable because of the distinctive ripples they produce. Fish predators like loons, grebes, and gulls cue in on these ripples and swim or fly toward them. When encountering a bait ball, a solitary loon will often hoot, alerting nearby loons, who quickly swim in the direction of the bait ball. Once there, loons and grebes dive beneath the bait ball, which results in many fish leaping out of the water and attracting gulls, who fly to the area to take part in the feeding. It is fairly common for large groups of ten to thirty loons to dive collectively in pursuit of these schooling forage species.

Loons in Lake Jocassee may forage differently from marine loons based on the vertical movements of baitfish. These movements are influenced by several environmental factors, such as degree of cloud cover, temperature, turbidity, and possibly wind speed and direction. When baitfish are deeper in the water column, they are harder to herd, so loons are more likely to switch to solitary foraging. We have observed solitary loons bring catfish and trout to the surface and swallow them whole. Given that catfish are predominately benthic feeders, this observation suggests at least some loons are diving between 150 and 200 feet to the bottom of the lake. The notion that loons are benthic feeders (for at least a portion of their time) is supported by Kevin Kenow, who has consistently found loons diving to the bottom of Lake Michigan, at depths of 150 feet or greater, during migration. Solitary loons will make dives of two to three (occasional four) minutes, duration, about the time necessary to reach the bottom of the lake and poke around for food.

Differences between Adult and Immature Loons

•

It is reasonable to expect juvenile or one-year-old loons to be less skilled at finding and catching prey in a new environment than experienced adults. In fact, inefficient foraging is cited as a contributor to elevated mortality of young birds. Many birds exhibit poorer foraging skills when they are immature. We speculated poor skills would cause juvenile loons to spend more time foraging than adult loons. When we examined this question in Louisiana, to our surprise, we found that adults foraged more than immatures. This counterintuitive result made us think about why this would be so. One key difference between adult and subadult loons during this time of year is that adults undergo a catastrophic wing molt, and immatures do not. So, adults require more energy from food to sustain and complete the growing of new flight feathers. Adults may also undergo hyperphagia (a drive to eat more than normal) prior to their long-distance migration, while immatures do not. However, I am curious to see if this pattern holds for loons in locations where migration distance is considerably less, such as the Maine coast.

Winter Mortality

•

According to survivorship data, more loons die in winter than any other time of the year. The leading cause of death during winter is emaciation syndrome. Birds with emaciation syndrome are thin and weak and eventually succumb and beach themselves. No single factor is responsible for emaciation syndrome; it can be brought about by a number of contributory causes, such as a high number of intestinal parasites, aspergillosis (a common fungal condition in wild birds), or simply malnourishment. Why do some loons experience malnourishment? For starters, storms stir up sediments, which increase water turbidity and reduce visibility, creating challenging conditions for foraging loons. In addition, storms may change or shift normal patterns of prey distribution and abundance, which likely impacts loon foraging success as well. Immature individuals of many animal species often live on the edge, and a storm may be a tipping point for many loons, especially young ones. Huge storms likely lead to an increase in emaciation syndrome. I have wondered if immediately after severe weather events, loons locally switch foraging strategies from catching fish to crabs. Crabs have higher salt levels than fish, which may induce some physiological stress related to a loon's need to eliminate excessive salt. Additionally, the blue crab is a major intermediate host for flukes, a primary wormlike parasite found in the guts of loons. Examination of emaciated loons from a large die-off in 1983 off the Gulf coast of Florida showed high concentrations of these flukes (average of 9,300/individual) in their intestinal tracts, suggesting that parasitic infections resulting from foraging on crabs may play a significant role in causing emaciation syndrome.

Working off the Louisiana coast, we came across two dead immature loons in 2011. They were emaciated, and necropsies showed that one died from aspergillosis, a fungal disease common in birds, and the other died from unknown causes. Though not a primary cause of death, aspergillosis usually gains hold as a secondary infection and can grow and lead to death in individuals who have compromised immune systems and/or nutritional deficiencies, both associated with stress from disease. In 2011, Dan Poleschook and Ginger Gumm observed a beached and dying adult loon on the Mississippi coast. When it expired, they took it to the Mississippi Research Station in Gulfport, where the resident pathologist determined the cause of death to be lead poisoning (the loon had a jig stuck in its bill). This was the first documented case of lead poisoning in a wintering loon in the ocean. Just how often this happens is unclear, but I have seen loons with fishing line and lures wrapped around their bills off a number of fishing piers in the southern United States. Further investigation is needed to document whether loons face a significant risk of lead toxicosis on their winter grounds.

Paul Spitzer may have been the first to suggest that marine biotoxins may be responsible for some winter loon mortality in the Gulf of Mexico. A marine biotoxin is a toxic substance produced by living organisms, such as red tide (algal blooms), from the organism *Alexandrium catenella,* a dinoflagellate. Many blooms occur with no visible bird mortality, and toxicity may vary within and between algal strains. Some algal toxins are neurotoxins that can impair a loon's foraging ability without immediate, acute toxicity, which may account for their emaciated appearance. We still have much to learn and understand about emaciation syndrome. Why, for example, do these large die-off events during the winter occur primarily in the Gulf of Mexico and on the Atlantic side of the Floridian coast but not in loon populations inhabiting northern regions, such as Puget Sound or New England?

Marine versus Freshwater Habitats

•

I have asked myself, "If I were a loon, would I be better off in the ocean or a lake?" I would likely choose the lake. A lake habitat would be less stressful because there would be fewer competitors and predators, less violent storms, less exposure risk to biotoxins (from marine algae), and no adjusting to added salinity in my diet. Yet, approximately 95 percent of all Common Loons winter in marine areas and have been doing so for hundreds of thousands of years. Clearly, they are remarkably adapted to living with these potentially adverse conditions and stresses. And who am I to underestimate the bounty of the ocean? Unless reservoirs and lakes have plenty of fish, loons may be able to meet their daily caloric needs more easily in the ocean than in lakes. Yet I wonder if loons fare better wintering on freshwater than marine habitats. Higher adult survivorship, especially in an animal that is long-lived, will usually lead to a population increase. For example, the snow goose population in North America exploded when adult winter survivorship increased. It increased because waterfowl management augmented the goose's winter cover and food. Can stocking reservoirs with fish lead to an increase in loon winter survivorship and, ultimately, a population increase?

The North American loon population has increased or stayed the same for the last couple of decades, despite an increase in factors that mostly lead to lower reproductive success and survival. There has been an increase in a major predator, the Bald Eagle, human recreational development and pressure remain high, mercury deposition has lowered loon productivity, and the use of lead tackle has led to adult mortality. The predominant explanation for the population's stability is increased management efforts such as the deployment of floating rafts adopted by loons for nesting, but I wonder if it isn't also the increased use of reservoirs leading to higher adult survival that is offsetting any loss of adults to a myriad of environmental and biological factors.

We have come a long way in our understanding of the behavioral ecology of wintering loons. Selection has shaped their distribution along the coasts as well as their foraging and social behavior. They are highly adaptable, capable of residing near shore or dozens of miles out to sea. They forage in shallow or deep water, solitarily or in groups. They eat fish of various sizes that live throughout the water column, and do not hesitate to eat invertebrates. They exhibit winter site fidelity and use the same general area annually, though young loons may switch. Loons are long-lived, and they show behavioral flexibility depending on local environmental and biological factors and can adapt their feeding strategies to maximize their survival. In spite of all of this, they still face a number of environmental threats on both their breeding and wintering grounds.

FURTHER READING

Alexander, L. L. 1991. Pattern of mortality among Common Loons wintering in the northeastern Gulf of Mexico. *Florida Field Naturalist* 19: 73–79.

Barr, J. F. 1996. Aspects of Common Loon (*Gavia immer*) feeding biology on its breeding ground. *Hydrobiologia* 321: 119–44.

Daub, B. C. 1989. Behavior of Common Loons in winter. *Journal of Field Ornithology* 60: 305–11.

Ford, T. B., and J. A. Geig. 1995. Winter behavior of the Common Loon. *Journal of Field Ornithology* 66 (1): 22–29.

Forrester, D. J., W. R. Davidson, R. E. Lange, R. K. Stroud, L. L. Alexander, J. C. Franson, S. D. Haseltine, R. C. Littell, and S. A. Nesbitt. 1997. Winter mortality of Common Loons in Florida coastal waters. *Journal of Wildlife Diseases* 33 (4): 833–47.

Haney, J. C. 1990. Winter habitat of Common Loons on the continental shelf of the southeastern United States. *Wilson Bulletin* 102: 253–63.

Johnson, J. E. 1971. Maturity and fecundity of threadfin shad, *Dorosoma petenense,* in Central Arizona Reservoirs. *Transactions of the American Fisheries Society* 100: 74–85.

Kenow, K. P., P. D. Adams, N. Schoch, D. C. Evers, W. Hanson, D. Yates, L. Savoy, et al. 2009. Migrations patterns and wintering range of Common Loons breeding in the northeastern United States. *Waterbirds* 32: 234–47.

Kenow, K. P., M. W. Meyer, D. C. Evers, D. C. Douglas, and J. Hines. 2002. Use of satellite telemetry to identify Common Loon migrations routes, staging areas and wintering range. *Waterbirds* 25: 449–58.

Lee, D. S. 1987. Common Loons wintering in offshore waters. *Chat* 51: 40–42.

Long, D., IV, and J. D. Paruk. 2014. Unusually large wintering foraging flock of Common Loons in the Gulf of Mexico. *Southeast Naturalist* 13 (4): 49–51.

McIntyre, J. W. 1988. *The Common Loon: Spirit of Northern Lakes.* Minneapolis: University of Minnesota Press.

Murphy, M. E., and J. R. King. 1992. Energy and nutrient use during moult by White-crowned Sparrow. *Ornis Scandinavica* 23: 304–13.

Paruk, J. D., M. Chickering, D. Long IV, H. Uher-Koch, A. East, E. A. Adams, K. A. Kovach, and D. C. Evers. 2015. Winter site fidelity in Common Loons across North America. *The Condor* 117: 485–93.

Paruk, J. D., D. C. Evers, J. W. McIntyre, J. F. Barr, J. N. Mager, and W. H. Piper. 2021. Common Loon (*Gavia immer*), version 2.0, *The Birds of North America,* edited by P. G. Rodewald. Ithaca, N.Y.: Cornell Lab of Ornithology. https://doi.org/10.2173/bna.

Paruk, J. D., D. Long IV, S. L. Ford, and D. C. Evers. 2014. Common Loons wintering off Louisiana coast tracked to Saskatchewan during the breeding season. *Waterbirds* 37: 47–52.

Pokras, M. A., and R. Chafel. 1992. Lead toxicosis from ingested fishing sinkers in adult Common Loons (*Gavia immer*) in New England. *Journal of Zoological Wildlife Medicine* 23: 92–97.

Sidor, I. F., M. A. Pokras, A. R. Major, R. H. Poppenga, K. M. Taylor, and R. M. Miconi. 2003. Mortality of Common Loons in New England, 1987–2000. *Journal of Wildlife Diseases* 39 (2): 306–15.

Spitzer, P. R. 1995. Common Loon mortality in marine habitats. *Environmental Review* (Canada) 3: 1–7.

Saving the Loons We Love

Conservation Threats

"Help! I need some help here!" I shouted. I could not pull the loon closer to the boat—it was a large male, very heavy, and I was far too extended and losing whatever leverage I had. The loon was in a net, and I was using all my strength just to keep it up and out of the water. If I lowered the net any farther, the loon would swim out. I did not have the strength to pull the bird toward the boat. Recall, a male loon from New England is one the heaviest loons on the planet, possibly weighing over 14 pounds. We had reached a stalemate, and the loon hung in the net a few feet above the water. I called again for some assistance. We were on Jenny Lake in southeastern New Hampshire, at about one o'clock in the morning, doing an important follow-up study examining mercury levels in loons. I needed the driver of the boat to reduce the distance between us or someone in the boat to assist me by grabbing the net handle, or this loon was going to escape. I was working with people I had just met, and most were not experienced loon catchers. Help seemed slow in coming, and my strength was giving out. In a flash, I thought of my two surgeries from years past—the first was to repair a torn rotator cuff in my shoulder, and the second to fuse my cervical spine to repair a severely ruptured disc—both of which had left my upper body weakened. I waited in pain for what seemed like an eternity.

Mercury

•

Dave Evers and his expert crew from the Biodiversity Research Institute are extremely skilled in the fieldwork of catching loons, drawing blood, snipping a couple of wing feathers, taking measurements, and banding and then releasing them. In 1990, Dave began collecting blood and feather samples from Common Loons in Michigan so they could be analyzed for mercury. His initial results from the eastern Upper Peninsula revealed that males consistently had higher mercury levels than females. He expanded his study to the western Upper Peninsula and Wisconsin and discovered that mercury levels in loons varied depending on location. Those early results spurred him to expand his sampling to loons across North America. Dave used the data in his doctoral research at the University of Minnesota. He found that mercury levels in loons varied across the continent, with some areas having significantly more; he designated those areas as hot spots. For example, the seacoast region of New Hampshire had individual loons on Jenny, Bow, and Swains Lakes with the highest concentrations of mercury on the continent. Loons turned out to be a perfect model to assess mercury levels in lake environments because they feed high on the food chain and the mercury bioaccumulates in their bodies. In other words, loons are biosentinels; their population and status can help monitor the health of the environment. The Common Loon would eventually become the avian poster child for mercury exposure, similar to how the Bald Eagle became a symbol of the dangers of DDT contamination.

Mercury affects the central nervous system and can impair behavior, attention span, muscle contraction, and in some cases survivorship. In the 1950s and 1960s, approximately six hundred people died and several thousand more became sick from eating fish and shellfish contaminated with mercury around Minamata Bay and Hyakken Harbor in Japan. The bay was contaminated by mercury-laden wastewater released by a local corporation. Mercury comes in different forms, inorganic and organic. The amalgam fillings found in your teeth and in thermometers are inorganic, which is biologically inert and for the most part unreactive. Some bacteria take the inorganic form of mercury and alter it chemically to become a biologically active form. This altered form can enter food chains, first in zooplankton, which are consumed by small fish. Larger fish, a major food source of loons, become contaminated when they eat the smaller fish. Along each food level, mercury bioaccumulates, becoming more concentrated and potentially more harmful. The mercury contamination in Japan made international headlines, and the dangerous effects of mercury exposure became known as Minamata Disease. Because of this event and other research, both the Food and Drug Administration and Environmental Protection Agency (EPA) advised women of child-bearing age, nursing mothers, and young children to avoid the consumption of top predatory fish, such as tuna, mackerel, and swordfish. They further recommended to limit consumption of tuna, for example, to 6 ounces per week. By the late twentieth century, it was clear that mercury can be harmful to many living species.

Mercury is a natural but uncommon component of the earth's crust. It gets released whenever volcanoes erupt and during mining, especially for gold. Mining operations in Ghana and Indonesia, for example, have released lots of mercury over the years. Once released, mercury gets absorbed by plants and becomes incorporated into fossil fuels and coal. When we burn coal or use fossil fuels, we release mercury into the atmosphere. Countries that use the most fossil fuels, including the United States, China, and India, have more issues with atmospheric mercury contamination. In the past 100 to 150 years, the amount of mercury in the atmosphere has risen tenfold, most of that coming from the burning of coal. Once released, it becomes airborne, traveling downwind and across state or national boundaries, even oceans. It eventually falls and enters watersheds, where it may get converted to the methylated, or toxic, form. Historically, mercury has been used in medical and dental offices, but its use has been widely curtailed because its toxic effects are now more widely known. Today, it is used primarily in fluorescent bulbs and street lamps.

Dave Evers, along with other researchers, such as Michael Meyer of the Wisconsin Department of Natural Resources and Neil Burgess from Environment Canada, expanded the connection between loons and mercury. Independently, they found that mercury was causing lower reproductive success in loons. The birds were less attentive both to incubating their eggs and providing food for their young (see chapter 5). Dave developed predictive models that give a good approximation of whether mercury concentration in a given blood or feather sample will have detrimental effects on a loon's reproductive success and how big any anticipated effect will be. Dave has dedicated much of his career to developing and establishing policies that monitor mercury pollution. With Senator Olympia Snowe from Maine, Dave tried introducing a national mercury-monitoring bill in Congress, but after several attempts, it never got off the floor. Undeterred, he expanded his mercury work internationally and presently works with the United Nations to oversee a strategic plan to reduce global mercury emissions.

How does all this relate to the loon in my net at Jenny Lake? The loon was important because nine to twelve years earlier, measures were taken to reduce mercury output in the seacoast region of New Hampshire, which had been identified as a hot spot. Legislation was adopted, and between 2001 and 2004 scrubbers were installed in smokestacks that burned coal in this region. The installment of scrubbers removed approximately 6,600 pounds of mercury from upwind incinerator emission sources. We hoped the loon in the net would give us some clues as to whether the removal of airborne mercury had had a measurable effect on the local loon population. A volunteer beside me rose to the occasion and helped me pull the loon across the water and into the boat. We took a blood and feather sample, released the loon, and months later got the results. What we found was a substantial decline of mercury in the blood of the Jenny Lake male and two other loons that our team caught that night. Our pilot study (by admission, an extremely small sample size of three individuals) suggested the scrubbers in the smokestacks were having the desirable effect. In 2011, the EPA passed the Mercury and Air Toxics Standards to decrease mercury deposition further across this region. Here is a case where the government's actions improved air quality and reduced the risk of mercury contamination to both wildlife and people. It was a win-win-win situation; however, the same agency has not felt the same way about lead, a metal that is equally, if not more, toxic and locally abundant in the environment.

Lead

•

In 2004, Dan Poleschook and his partner Ginger Gumm observed a female loon on South Twin Lake in Washington State die on its nest while incubating eggs. In 2012, Tiffany Grade observed a female loon on Squam Lake in New Hampshire get viciously attacked by another female loon to the point that it had to get up on shore to get away from the onslaught. In the end, she too died. The two loon deaths were linked, though separated geographically and temporally. Could these deaths have been avoided? Absolutely. In 2001, two researchers from the University of Wisconsin assisted Dave and me in catching loons in New

England to see if we could identify why so many loons were dying across the country. They brought a portable X-ray machine. When we caught a loon, we boated over to shore, where we positioned the loon under the machine long enough for the technician to snap some pictures. We were specifically looking for lead, because lead is toxic to wildlife in concentrated doses, such as those found in ammunition and fishing tackle. We found definitive evidence of lead tackle inside the loon.

Though it occurs naturally, lead has no biological function. Lead affects the brain and other major organs; the body can mistake it for other beneficial metals, such as zinc and copper. At high doses, lead affects nerve impulse transmission, leading to paralysis, then death. This is what happened to the loon that Dan and Ginger observed on South Twin Lake. At low to moderate concentrations, it leads to excessive dehydration, gastrointestinal problems, loss of weight, weakness, and fatigue. At high concentrations, it bursts red blood cells, leading to anemia and adding to fatigue already produced by loss of water. This is what happened to the loon that Tiffany observed being attacked by another loon; it was too weak to defend itself. Tiffany recovered the dead loon and took it to a wildlife clinic for examination. The necropsy results verified what she had suspected: lead poisoning. Tiffany had been watching this particular loon for years and could tell something was amiss—it was not acting normally. Besides showing fatigue and lethargy, loons suffering from lead toxicosis will droop a wing, exhibit excessive head shaking and gaping, and produce green feces. A single split shot sinker ingested contains enough lead to poison an adult loon and produce these symptoms, eventually killing it in roughly two to three weeks.

The World Health Organization estimates that 143,000 to 400,000 people die each year from ingestion of predominately lead paint. Many children in contact with lead paint will develop intellectual disabilities. Historically, lead was used in paint because it was an effective fungicide. Over time we have removed lead from gasoline, children's toys, and paint because of its documented toxic effects. One would think we would want to make the environment safe for wildlife, too. Hundreds of birds, including eagles, condors, waterfowl, cranes, doves, and loons, die annually from the ingestion of lead sinkers, jigs, and ammunition. Fishing lines break, and the sinker or split shot at the end of the line makes it way to the bottom. Once there, waterfowl or a loon may pick it

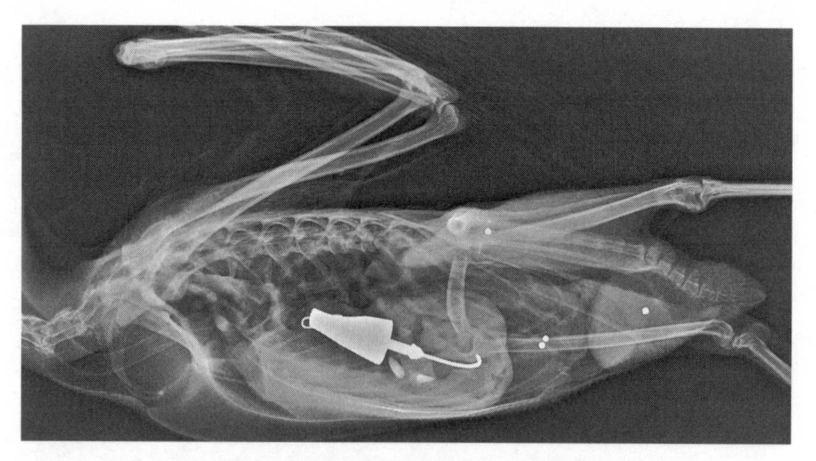

An X-ray image of the body of a Common Loon that died from ingesting lead fishing tackle. Courtesy of Mark Pokras.

up, most likely mistaking it for a pebble, which they use as grit in their gizzards to masticate food. Loons most often die from swallowing a jig head. On occasion, a fish breaks the line before it can be pulled in, and the jig that hooked the fish is still attached. The fish gets caught by a loon and is swallowed, jig and all. Sometimes, a loon becomes completely entangled in fishing line, and unless someone catches and removes it, it will likely starve or drown.

Mark Pokras, recently retired director of the Wildlife Clinic at Tufts University's Cummings School of Veterinary Medicine, has spent several decades documenting the effects of lead on loons. With his students and volunteers, he has performed more necropsies on loons than anyone else, more than 3,500. They determined the cause of death for all the loons collected that died in New England over a fourteen-year period, from 1987 to 2000. Pokras and his team found that 222 (44 percent) of the breeding adults died from lead toxicosis. Of the 522 Common Loon carcasses examined, 118 had ingested lead objects with the following frequency: sinkers, 48 percent; jigs, 19 percent; split shot, 12 percent; shotgun pellets and bullets, 11 percent; other lead, 8 percent; and unknown lead, 2 percent. Dan and Ginger collected twenty-seven loon carcasses in eastern Washington from 1996 to 2010 and were able to determine the cause of death for twenty-three of them. Thirteen of those loons (57 percent) died from fishing-related activities, with the leading cause being lead poisoning by ingestion of lead sinkers, split shots, and jigs (nine), followed by fish net entanglement (two), fish-hook puncture (one), and other metal (one). Lead clearly poses a major problem for loons and their continued survival.

Lead Legislation

A leading conservationist of his day, George Bird Grinnell recognized and wrote about the toxic effects of lead on birds as far back as 1894. He noted that ducks and geese were vulnerable to lead poisoning by swallowing shotgun pellets. It took a long time, but in 1991 the U.S. Fish and Wildlife Service passed an ordinance banning lead shot for waterfowl hunting in all fifty states. Canada followed suit in 1999. The reason behind these laws was declining waterfowl populations. Since the ban, waterfowl in both countries have rebounded. But fishing tackle is still a problem. The ingestion of lead tackle has led to the death of thousands of Sandhill Cranes, Trumpeter Swans, and Mute Swans. The problem was deemed critical in Great Britain, and the government banned the use of lead sinkers on fishing tackle altogether in 1987. The U.S. EPA attempted to pass a nationwide bill to ban lead sinkers in 1994, but it failed due to strong opposition by sportsmen's groups. In 1995, the U.S. Fish and Wildlife Service (USFWS) banned the use of lead sinkers in wildlife refuges where there was a reasonable chance of affecting Trumpeter Swans and Common Loons, including Red Rock Lakes National Wildlife Refuge (NWR) (Montana), National Elk NWR (Wyoming), and Seney NWR (Michigan), and in Yellowstone National Park. In 1997, Environment Canada and Parks Canada prohibited the possession of lead sinkers and jigs (1.8 ounce or less) in national wildlife areas and national parks.

In recent decades, the EPA and USFWS failed to pass further measures to reduce the use of lead in hunting and fishing gear despite mounting evidence of its toxicity in the environment. Several states took a leadership role to address the issue, and since 2000, six states have passed legislation curtailing the use and/or the sale of lead fishing tackle. New Hampshire has banned lead sinkers of 1 ounce or less, and jigs 1 inch or less statewide. Massachusetts has banned lead sinkers, but only on the two main reservoirs loons use for breeding, Wachusett and Quabbin. A few years later, Maine (2002), New York (2004), and Vermont (2007) banned the sale of lead sinkers of one-half ounce or less and restricted their use as well. The Washington Department of

Fisheries and Wildlife passed legislation in 2010 that partially banned the use of lead tackle on twelve of the thirteen lakes used by loons during the breeding season. Since 2010, no additional state has adopted any legislation on lead restrictions, though some of the original six states have made amendments to their original laws. For example, Massachusetts has since expanded its ban statewide to include all small lead fishing weights. Minnesota, the contiguous state with the most Common Loons, has run into fairly stiff opposition to passing regulatory legislation, despite also seeing adverse effects from lead on Bald Eagles and Trumpeter Swans.

Why would the EPA pass legislation to reduce the environmental risk of mercury but not lead, when both are equally toxic to wildlife? There are two fundamental differences between mercury and lead legislation. First, a huge lobbying group of sports enthusiasts opposes any legislation that will restrict or outlaw the use of lead fishing tackle. Mercury legislation faced no such opposition, though private industry was certainly unenthusiastic about adopting changes because of the high cost of installing scrubbers. Second, any adoption of lead legislation would mean that anglers would have to purchase nonlead tackle at their own expense. Mercury legislation required no personal cost nor change in human behavior. The position of the American Sportfishing Association (ASA), a powerful lobbying organization, is that lead tackle does kill eagles and loons but not a lot of them, or not enough to warrant a nationwide ban on lead. It argues that a nationwide ban on the selling of lead would cause economic hardships for many small businesses. The ASA believes it is better to adopt a partial ban on lead use in local lakes only when backed by science that reveals there is a serious problem. It further contends that if lead ingestion is having such a dire effect on eagle and loon numbers, why do their populations appear to be stable and even growing in some places?

Tiffany Grade and her colleagues at the Loon Preservation Committee (LPC), the oldest nonprofit loon conservation group in the United States, have data that show that loon mortality due to lead toxicity is affecting populations in New Hampshire and that the only reason loon numbers are not down further in New Hampshire is because of the intensive management efforts annually performed by LPC staff. Although both positions cannot be right, it is conceivable that population level effects may be occurring locally, not nationally. But there have been alternatives to lead fishing tackle for years, such as bismuth, steel, tin, and tungsten. The cost to switch out one's lead tackle to a nonlead alternative will cost money and could be an inconvenience, for sure, but compared to the expense of one fishing trip, the cost to switch out tackle is neither exorbitant nor prohibitive. Granted, economic forces are at play as well, and small fishing shops would likely be most affected by a nationwide ban. That is why people like Harry Vogel, the director of the LPC, feels a gradual shift from lead to nonlead alternatives over several years will help everyone, including manufacturers and retailers, in making the needed adjustments.

I am left wondering how many more eagles, loons, condors, cranes, and swans have to die before a change in policy is warranted. The EPA was designed to protect the health of humans and the environment by writing policy and enforcing regulations passed by Congress. The EPA enforces a number of laws protecting people from the harmful effects of lead, such as the Toxic Substances Control Act, the Clean Water Act, the Clean Air Act, and the Safe Drinking Water Act. In addition, the agency conducts environmental research and is responsible for assessment. But the EPA's assessment of lead toxicosis in the environment appears to be in line with the ASA position, acknowledging that lead is toxic to wildlife but not enough to warrant a national ban of lead tackle. Despite the mounting evidence that ingested fishing tackle leads to numerous wildlife deaths, the EPA refuses to recognize that there is a serious problem nationally. Because the federal government has failed to act on banning lead tackle, the responsibility shifts to state governments to take action.

Dan, Ginger, and Tiffany had spent more than seven years observing and monitoring each of those loons that died from lead toxicosis. They knew the birds' life histories (whom they mated with and how many chicks they produced each year), but more important, they knew them as individuals. One cannot help but to get to know an animal one observes for seven years. I am sure all pet owners can relate. These researchers understood the loons' personal quirks and the oddities that made them unique, and it crushed their spirits to watch them die. Some deaths are accidental, and some are just inherent to being a wild animal living with risks amid the elements, but some fall under our watch and are directly caused by our actions or inactions. Those deaths are hard to stomach.

Oil Spills

•

In 2012, I was the principal investigator of an Earthwatch expedition studying the impacts of the Deepwater Horizon oil spill on migrant Common Loons. My co-principal investigator, Darwin Long IV, and I drove down the Mississippi Delta to Port Sulphur, where we met the volunteers. After breakfast the next morning, I went over the day's objectives. We had rented two boats for the day to survey the area for loons and make behavioral observations. We met our boat captains, Todd Seither and Jay Winters. We split the eight volunteers, and half went with Darwin and Todd, and other half with Jay and myself. The groups were excited about their first day on the water. They signed up for the expedition because they had all read and heard so much about the environmental disaster in the news. The weather was cooperative, mild and sunny, and birdlife was abundant with Brown Pelicans, Laughing Gulls, Royal Terns, Glossy Ibis, and Snowy Egrets. We boated out among channels that would lead us to more open water, marveling at all the pelicans, egrets, ducks, gulls, and terns along the way. I had to remind everyone we were looking for loons. Our boat was passing through the channel first, with Darwin's behind, when we noticed something unusual off the stern. It was an adult dolphin, but it was acting oddly. It was not diving regularly but stayed near the surface, and appeared to be pushing something. We were too far to tell, even with binoculars, so we motored closer. None of us was mentally prepared for what we observed next. We could see it was a calf, but when we drew closer, we saw that the young one was not moving, it was dead, and the mother was trying to keep it afloat. The devastation to marine wildlife caused by the oil spill made us quickly remember why we were there.

On January 19, 1996, the oil tanker *North Cape* caught fire and grounded off the coast of Rhode Island, spilling approximately 828,000 gallons of home heating oil into the Atlantic Ocean. The spill resulted in the death of nearly 2,300 birds, including a projected 402 Common Loons. On Sunday, April 27, 2003, Bouchard Transportation Company Barge 120 ruptured a cargo tank, spilling approximately 98,000 gallons of fuel into Buzzards Bay, Massachusetts. The oil that was released impacted more than 53 miles of shoreline, killing several hundred birds, roughly 50 of them loons. Waterbirds like loons, scoters, and eiders remain watertight because their feathers are zipped and interlocked in a way that keeps water away from their skin, comparable to a human donning a wet suit (see chapter 3). I know from personal experience that a leak in a dry suit is quickly noticed because they are designed to keep water out. Once water enters the dry suit, it is just a matter of time before one gets cold. In waterbirds, oil damages and interferes with the way feathers overlap and interlock, disrupting the waterproof barrier. Birds with oiled plumage are robbed of heat because they are exposed to the cold ocean water. Heat loss leads to increased metabolic rates, which causes rapid metabolism of body fat, further stressing the individual. Because the plumage no longer repels water but retains it instead, it adds excessive weight, resulting in some cases in the bird drowning. However, most oiled birds die from hypothermia. How much of the plumage needs to be oiled before a waterbird dies from exposure? The answer depends on the species in question and environmental conditions. Loons are considered highly sensitive and vulnerable to oiling because they depend on their watertight plumage for body temperature regulation (in this way, they are no different from other obligate waterbirds). Therefore, loon mortality will be greater if loons are exposed to oil during the winter compared to the summer. Just a trace amount of oil (for example, less than 5 percent of the body surface) present on a loon during the winter off the mid-north Atlantic coast, may be enough to cause its death. Marine birds in warmer ocean waters, like the Gulf of Mexico, may survive trace oiling, but light (6 to 20 percent of the body surface) and moderate oiling (21 to 40 percent) may lead to mortality rates of 80 to 90 percent and 90 to 100 percent, respectively.

To learn how to mitigate the impacts of oil spills, we need to know not only the number of loons that died from direct contact with the oil but also how many died indirectly, through chronic exposure. Research suggests that more marine birds die from the chronic effects of oil exposure than from the initial acute exposure. We also need to know how these deaths influence the overall population. The death of an adult from an oil spill will have a greater impact on the population than the death of an immature loon. Not all immature loons will live to breeding age, but once an adult is eliminated, all the future young it could have produced over its lifetime are gone, never able to contribute to future generations. At this point, we can ask how many breeding adult years were lost. Models estimate such numbers, and in the *North Cape* oil spill, 402 loons died, resulting in between 5,028 and 5,628 lost loon years. To compensate for lost loon years, one mitigation plan would be to protect or establish loon nests, deploy artificial nesting platforms, and monitor them. The party responsible for the spill would pay for those costs. Litigation involving marine oils spills is a long, drawn-out process, which often takes ten or more years to reach settlements. Unfortunately, if recent history is any indicator, the question is not if another marine oil spill will occur but when and where.

The Deepwater Horizon Oil Spill

In May 2010, folks at the Sigurd Olson Environmental Institute (SOEI) at Northland College were raising some important questions: How would the Deepwater Horizon oil spill impact wintering loons, and what, if any, effect would it have on the state's or region's breeding population? Deepwater Horizon was the largest industrial oil spill in the world's history. Beginning in April 2010, over 200 million gallons of oil leaked before the well, about 50 miles from the coast in the Gulf of Mexico and roughly over a mile below the surface, was sealed sometime in July, three months later. Stretching over 600 miles, the oil slick reached the coast in August, predominately hitting Louisiana but also impacting Alabama, Mississippi, and parts of Florida. Solid data linked loons breeding in the upper Midwest states of Michigan, Wisconsin, and Minnesota to wintering sites in the northern Gulf of Mexico. Loon enthusiasts and researchers were naturally concerned.

The mission of LoonWatch, a program of the SOEI, is to protect Common Loons and their aquatic habitats through education, monitoring, and research. I served on its research board for five years. That May, when folks met at the SOEI, my office was located only yards away, but I was busy teaching an intensive four-week-long field ornithology course for Northland College. The staff and interested parties who showed up that day asked who was going to lead this important research. A branch of the federal government that deals with natural resource damage assessments (NRDA) was looking for a group with the expertise to conduct the assessment on waterbirds that use open water, such as loons, pelicans, and gannets. That summer, Dave Evers, executive director at Biodiversity Research Institute (BRI), in Maine, and his versatile and talented team of biologists received several grants from the NRDA program to assess the impact of the Deepwater Horizon spill on waterbirds, Common Loons being but one of several target species.

In July, Dave called me to see if I was interested in heading up the loon assessment part of the grant. I told him that it was a unique opportunity, but that I was settled as a full-time, tenured biology faculty member at Northland College. I couldn't just pick up and leave. Dave was persistent and asked me again two weeks later. We discussed the job in more detail, and I talked with my wife and close friends about leaving Northland for this new opportunity. It was a difficult decision. We loved our community, our friends, and the outdoors of the region. The students at Northland were passionate about sustainability and saving the earth and its inhabitants. I also had a good research program going. I was forty-eight years old and comfortable . . . why leave? This is why: Deepwater Horizon was the world's largest marine oil spill in the history of the petroleum industry. Our findings and their implications could influence policy and regulations for decades. I could be involved in shaping future environmental assessments of oil spills and their regulation and policy. I had an opportunity to experience something novel and to grow professionally. After much consideration, I accepted Dave's offer. That December, my wife and two daughters and I drove from Wisconsin to Maine, and I started my new position at BRI.

I hit the ground running with almost no time to breathe. I was thrust immediately into hiring and overseeing a team of biologists to conduct the fieldwork. I immediately thought of Andrew East, an exceptional student I had at Northland College. I tracked him down and offered him the job. The next fire that needed attending to was our assessment plan. This was more complicated, as it involved multiple parties and had to be arranged by conference calls. After too much deliberation for my taste, the plan was adopted and we were up and running.

Our role in the assessment consisted of conducting loon surveys by boat and land over the course of one month (February 17–March 15), from Terrebonne Bay, Louisiana, to Apalachicola, Florida, the region most impacted by the spill. For each loon observed, we would assess if it was oiled and, if so, how much of the body was affected. For example, levels were categorized as None, Trace (greater than 0 but less than 5 percent of the bird's body surface), Light (6 to 20 percent), Moderate (21 to 40 percent), and Heavy (more than 40 percent). Oiling rates were estimated for birds only when more than 50 percent of the body surface could be clearly observed. Each bird that was clearly observed was assigned an oil exposure category based on visual and photographic inspection. Darwin Long IV would head up and supervise the field crews. Working with someone I knew and trusted was a huge comfort to me. Other biologists were assessing different species, such as egrets and pelicans. I flew to Louisiana that winter and met with all the field crews. Seeing the shoreline covered with sludge and oil was hard. British Petroleum, the company responsible for the accident, had hired hundreds of people, who were busy "cleaning" the spill in the area. I was devastated as I watched a Great Egret fly overhead with oil on its plumage.

By the end of the study we had observed more than one thousand loons but were able to assess only about one in ten of them (about 9 percent) because we could rarely see their undersides. Out of the loons assessed, roughly one in four (about 23 percent) had evidence of oil on its body. We found less oil than we had feared. Nearly 75 percent of the loons with oil had trace amounts (a speck on the belly, the bill, the webbing), and almost 25 percent had light levels. Only one loon (4 percent) was assessed in the moderate category. Some of the oil had sunk below the surface and settled on the ocean floor. We surmised that loons picked up oil when they foraged on the bottom, burrowing for crabs and bottom-feeding fish.

Surveys were also being done for other types of wildlife. If any party came across a dead bird, it was either picked up, or marked and later picked up by someone else. In the first year following the oil spill, seventy-five loons were collected and turned in to the U.S. Fish and Wildlife Service. Most birds that die from direct oiling are never collected: they simply drown, drift out to sea, or are eaten. So what proportion of the dead birds are collected? This question is not easily answered. It will vary by species (size, shape, where they first contacted the oil, their ability to fly or swim), by type of shoreline (can it be easily surveyed, is there suitable access), by environmental conditions such as currents and wind, and by the amount of effort devoted to finding carcasses. A conservative estimate is one in twenty marine birds that die from direct oil exposure is recovered, but it was reported that about nine hundred loons were killed directly. Using both the frequency and oiling levels (trace, light, or moderately oiled) that we observed, we calculated mortality rates and estimated as much as 4 to 11 percent of the wintering population in the northern Gulf of Mexico could have died from the Deepwater Horizon oil spill.

Suddenly, almost before we knew it, our assessment phase was over. There would be no more surveys or any additional research that would involve long-term monitoring. There would be no funding to determine if loons were exposed to the oil either through direct contact or indirectly through the food chain. There would be no tracking of loons to link where these exposed birds breed. It was frustrating and disheartening. I uprooted my family and gave up a job I loved for one month of fieldwork. Seriously? The project was supposed to last a couple of years. Dave assured me that there was money to keep me on for a while. There were data to analyze and reports to write, which would keep me busy for another six months, but the new reality was if I wanted to stay employed there, I would have to write grants to support my research. If I wanted to keep investigating the effects of Deepwater Horizon on loons, I would have to find my own funding. I called Earthwatch Institute and pitched my project. It turned out to be a perfect union: they would support my research financially and provide volunteers to assist in the work, and I would get to keep on doing what I loved. By the following winter, in 2012, I had secure funding and forty-four volunteers helping me gather data, catch loons, and conduct behavioral research in the coastal Louisiana towns of Port Sulphur and Buras, both areas that were moderately to heavily impacted by the spill.

Oil is highly toxic and is made up of several classes of hydrocarbons. One class, polycyclic aromatic hydrocarbons (PAHs), are especially toxic to wildlife and people. Whereas mercury and lead are toxic at concentrations of parts per million, PAHs are toxic in parts per billion (ppb). They initiate cancers and tumors, suppress the immune system, damage red blood cells and major organs like the liver and kidneys, disrupt hormone production, impact gut absorption, and affect the body's ability to regulate salt. I read about a study that used PAHs as an assessment tool for a marine oil spill that took place off the coast of Spain. PAHs are not ideal to work with as an assessment tool because one needs to obtain a blood sample for analysis and that requires catching the bird. In addition, because the livers of birds and other vertebrates can break down PAHs, it remains unclear how well measuring them works as an assessment tool. The study in Spain produced positive results, so I decided to adopt the same protocols they used, but that meant I needed to catch loons in the winter, which is a difficult task.

We used the same night-capture method that was used at Walker Lake and Morro Bay, with the main team consisting of a driver, a spotter, and a netter. Once we landed a loon, other people could help hold it and record data. We drew blood and took a feather sample from each bird and had them analyzed for PAH. We also performed a standard blood test that measures glucose levels, the ratio of white to red blood cells, types and number of white blood cells, and so on. To ensure success, we needed a competent and knowledgeable boat captain. Todd Seither knew the local area well but was more interested in fish than birds initially. He did not know what a loon looked like, nor did he seem to want to learn. But those early days of indifference were short-lived. He became an integral member of our capture team and was the first one in the boat to spot a loon. Going into this, I knew we were going to have difficulties catching wintering loons, but I remained hopeful, knowing that we had caught them in Morro Bay. I did not think we would have to worry about fog, but some nights it would roll in quickly, leaving us absolutely blind in the water and having to rely on Todd's GPS to trace our route. The waters of coastal Louisiana contain many obstacles, including shallow areas, which needed to be avoided. Arguably a higher risk in foggy conditions was colliding with another vessel. A stream of boat traffic was shuttling workers from the coast to the oil platforms, and they would not expect to come across a 24-foot boat at midnight in the middle of the ocean. At sundown the dampness and cold temperatures chilled us to our cores. But on milder nights, the stars were breathtaking, and we would shut the motor off and just gaze upward.

In January, we started each trip when it was dark, around 6:00 p.m., and ended around midnight, but as daylight increased, we would start a bit later and come in later. Once we had been out for eight hours and had caught three loons. It was around 2:00 a.m., and we were spent and tired. Todd turned on the radio to "Margaritaville," cranked the volume, and each of us sang as if we were auditioning for *America's Got Talent*. If any loons were within earshot, they would have noted that we each sang in our own key and pitch, but the volume was the same. We did not win any awards that night but were satisfied with the knowledge that we had put in a good night of work. The shared smiles made our tired bodies lighter and gave us the needed energy to unload the boat, drive home, and work up the field samples. Beginning in the winter of 2011 and during the next six years, we caught 128 loons, plus 10 additional recaptures, averaging roughly 20 loons a year. The work could not have been done without the help of the Earthwatch volunteers and invaluable assistants, such as Darwin Long IV, Andrew East, Kristin Kovach, Allie Byrd, Alex Dalton, and especially the well-organized and reliable Hannah Uher-Koch (who stayed up frequently until 4:00 a.m. working up samples). I have been fortunate to work with several highly talented and devoted biologists during my years and am eternally grateful for all their hard work and dedication. They are a source of hope for a better tomorrow.

The entire investigation lasted seven years. During the first two winters (2011 and 2012), we found many immature loons on retention ponds and dikes along the road on the peninsula, similar to places ducks are found. I have never witnessed this behavior in loons and thought it highly peculiar. As we watched these birds, many appeared weak and emaciated. One loon was so weak while swimming on a retention pond that Darwin was able to grab it by the neck and lift it out of the water. We administered Pedialyte® to it and to other loons we captured that appeared unhealthy. I think the loons had moved from the coastal estuary to the landlocked waterbodies in search of food. We caught more loons that second winter than any other year, probably because the loons came closer to shore to find food (they were also still common in the retention ponds and roadside dikes). During our third and every subsequent year thereafter, we never observed immature loons in the retention ponds or dikes, although we still commonly observed them within marinas and near shore.

The year with both the highest percentage of loons with PAHs and with the highest PAH concentration was not year one or two but year three. The PAH frequencies and average concentrations for the first three years looked like this: frequencies (0 percent, 24 percent, 62 percent) and concentrations (0, 6, and 104 ppb). What could have been responsible for such a drastic increase in both frequency and concentrations of PAHs in loons? Normally contaminants move up the food chain, and most invertebrates cannot detoxify PAH like vertebrates can. Thus, a possible reason is that it took several years for PAHs to move up through the food chain. However, if loons are feeding on crabs, which are fairly low on the food chain, any contaminant, given the higher temperatures in the Gulf and how fast chemicals degrade or move through the system, should have been be detected in one year. But in year three,

Hurricane Isaac happened. For several days in late August 2012, high winds of 60 to 85 miles per hour occurred over our study area in southeast Louisiana. My conclusion is that oil from the spill had settled on the ocean bottom and was stirred up by the hurricane-force winds. The contaminants in the oil then entered the food chain. We wondered what would happen in year four. Would loons show signs of exposure and, if so, at what concentrations?

We caught only thirteen loons in year four (2014), and not a single one showed evidence of having been exposed to PAHs. The spike had run its course in one year, and this raised the prevailing question: Was the worst over? I was not sure what to expect in year five, so when the results came back I had my answer. The worst was certainly not over. Just like in year three, loons caught in year five had both high frequency (54 percent) and high concentrations (average of 90 ppb) of PAHs. Hoping to determine where these PAHs came from, I asked the lab I was working with if it could trace the oil in the loon blood back to the Deepwater Horizon spill, and they told me that unfortunately they could not. The frequency of PAHs in loon blood remained high, 55 percent and 53 percent, in both years six and seven. Also, the PAH concentrations in loons during year six (18 ppb) and year seven (48 ppb) were still much higher than in the first two years of the study. So, seven years after the oil spill, roughly half of the loons my team caught had PAHs in their blood, and worse, the concentrations were still fairly high.

But were the PAHs affecting these loons, and if so, how? I found that increasing concentrations of PAHs in loons affected their weight. Loons with higher PAH concentrations weighed less than loons without PAHs and than those that had lower PAH concentrations in their blood. PAHs affect the gut lining and likely a loon's ability to absorb food. This would alter their behavior and activity level, causing them to forage more than normal. If they could not acquire enough food, their ability to put on the requisite fat needed prior to migration would be compromised. PAHs have been shown to lower red blood cell numbers, which is critical for delivering oxygen and removing carbon dioxide from cells. Low red blood cell counts result in anemia, causing fatigue. Several loons we found were anemic (about 6 percent), and these individuals weighed significantly less than nonanemic loons did. Also, loons with PAH detections had 3 percent lower levels of red blood cells than loons without any PAH detections. Finally, I found evidence of chronic inflammation (elevated white blood cells) in Common Loons with PAH exposure. The health and condition of a loon prior to migration are critical to a successful journey. It is all connected. Anything that impacts a loon's overall health and condition in the winter is going to impact its ability to migrate and breed successfully.

The accumulative effects of indirect oil exposure through the food chain likely had an impact on both the wintering population of loons in the Gulf and, potentially, the breeding population where they migrated. However, the NRDA plan for loons did not include any future monitoring of long-distance migrants leaving the Gulf. My recommendation for future oil spill assessments would be to allocate more time and resources to document the sublethal effects of oiling on wildlife. There are no studies on immune suppression in birds from oiling, for example. I carried out an immunosuppression pilot study with Keith Grassman, an immunotoxicologist at Grand Valley State University. Our preliminary data showed that several loons exposed to oil may have had suppressed immune systems. More studies designed to measure cellular and organ damage in birds would be useful. We also need to make sure the connection between wintering and breeding species is established and that the breeding population is assessed as well. Did birds exposed to oil have lower reproductive success or survival than ones not exposed? No money had been set aside in the case of Deepwater Horizon for this type of assessment in loons. Since the assessment phase is generally done in the first couple of years, it is important to also ensure that funds are reserved for later evaluative assessments.

I wish I knew if PAHs were present in my study area before I started testing for them in 2011, but the fact that we detected only extremely low concentrations during our first year, nine to eleven months after the spill, suggests that if they were present, they were in low concentrations. Field studies following a marine oil spill are challenging because pre-spill data are often lacking. I could not raise funds to support my loon research in Louisiana in 2018 and did not attempt to obtain any for 2019 or 2020, but I hope to acquire funding for 2021, our study's ten-year anniversary. British Petroleum agreed to an overall environmental settlement of 18 billion dollars, the largest in history, with most of that money going to the Gulf States. Minnesota received 7.2 million dollars from the Deepwater Horizon settlement for Common Loon restoration. The restoration money can be used to buy conservation easements and shorelines with critical loon habitat to ensure they have quality nesting areas. It could also be put toward educating anglers about the toxic effect of lead tackle on loons and other wildlife. Marine oil spills are not going away: we have to be ready for the next one.

Further Reading

Burgess, N. M., and M. W. Meyer. 2008. Methylmercury exposure associated with reduced productivity in Common Loons. *Ecotoxicology* 17 (2): 83–91.

Evers, D. C. 2006. Loons as indicators of aquatic integrity. *Environmental Bioindicators* 1: 18–21.

Evers, D. C., Y. J. Han, C. T. Driscoll, N. C. Kamman, M. W. Goodale, K. F. Lambert, T. M. Holsen, C. Y. Chen, T. A. Clair, and T. Butler. 2007. Biological mercury hotspots in the northeastern United States and southeastern Canada. *Bioscience* 57 (1): 29–43.

Evers, D. C., J. D. Kaplan, M. W. Meyer, P. S. Reaman, W. E. Braselton, A. Major, N. Burgess, and A. M. Scheuhammer. 1998. Geographic trend in mercury measured in Common Loon feathers and blood. *Environmental Toxicology and Chemistry* 17: 173–83.

Evers, D. C., L. J. Savoy, C. R. DeSorbo, D. E. Yates, W. Hanson, K. M. Taylor, L. S. Siegel, et al. 2008. Adverse effects from environmental mercury loads on breeding Common Loons. *Ecotoxicology* 17 (2): 69–81.

Evers, D. C., M. Sperduto, C. E. Gray, J. D. Paruk, and K. M. Taylor. 2019. Restoration of Common Loons following the North Cape Oil Spill, Rhode Island. *Science and the Total Environment* 695: 133849.

Grade, T. J., M. A. Pokras, E. M. Laflamme, and H. S. Vogel. 2018. Population-level effects of lead fishing tackle on Common Loons. *Journal of Wildlife Management* 82: 155–64.

Hernberg, S. 2000. Lead poisoning in a historical perspective. *American Journal of Industrial Medicine* 38: 244–54.

McIntyre, J. W. 1988. *The Common Loon: Spirit of Northern Lakes.* Minneapolis: University of Minnesota Press.

Meyer, M. W., D. C. Evers, J. J. Hartigan, and P. S. Rasmussen. 1998. Patterns of Common Loon (*Gavia immer*) mercury exposure, reproduction and survival in Wisconsin, USA. *Environmental Toxicology and Chemistry* 17 (2): 184–90.

Nocera, J. J., and P. D. Taylor. 1998. In situ behavioral response of Common Loons associated with elevated mercury (Hg) exposure. *Conservation Ecology* 2 (2): 10.

Paruk, J. D., E. M. Adams, H. Uher-Koch, K. A. Kovach, D. Long IV, C. Perkins, N. Schoch, and D. C. Evers. 2016. Polycyclic aromatic hydrocarbons in blood related to lower body mass in Common Loons. *The Science of the Total Environment* 565: 360–68.

Paruk, J. D., D. C. Evers, J. W. McIntyre, J. F. Barr, J. N. Mager, and W. H. Piper. 2021. Common Loon (*Gavia immer*), version 2.0. *The Birds of North America,* edited by P. G. Rodewald. Ithaca, N.Y.: Cornell Lab of Ornithology. https://doi.org/10.2173/bna.

Paruk, J. D., D. Long IV, S. L. Ford, and D. C. Evers. 2014. Common Loons wintering off Louisiana coast tracked to Saskatchewan during the breeding season. *Waterbirds* 37: 47–52.

Paruk, J. D., D. Long IV, C. Perkins, A. East, B. J. Sigel, and D. C. Evers. 2014. Polycyclic aromatic hydrocarbons detected in Common Loons wintering off coastal Louisiana. *Waterbirds* 37: 85–93.

Paruk, J. D., I. J. Stenhouse, B. J. Sigel, E. M. Adams, W. A. Montevecchi, D. C. Evers, A. T. Gilbert, M. Duron, D. Long IV, J. Hemming, and P. Tuttle. 2020. Oiling of American White Pelicans, Common Loons and Northern Gannets in the winter following the Deepwater Horizon (MC252) oil spill. *Environmental Monitoring and Assessment* 191: 817. https://doi.org/10.1007/s10661-019-7925-y.

Pokras, M. A., and R. Chafel. 1992. Lead toxicosis from ingested fishing sinkers in adult Common Loons (*Gavia immer*) in New England. *Journal of Zoological Wildlife Medicine* 23: 92–97.

Pokras, M. A., S. Rohrbach, C. Press, R. Chafel, C. Perry, and J. Burger. 1993. Environmental pathology of 124 Common Loons from the northeastern United States. In *The Loon and Its Ecosystem: Status, Management, and Environmental Concerns.* 1992 American Loon Conference Proceedings, edited by L. Morse, S. Stockwell, and M. Pokras, 20–53. U.S. Fish and Wildlife Service, Concord, N.H.

Sidor, I. F., M. A. Pokras, A. R. Major, R. H. Poppenga, K. M. Taylor, and R. M. Miconi. 2003. Mortality of Common Loons in New England, 1987–2000. *Journal of Wildlife Diseases* 39 (2): 306–15.

Stone, W. B., and J. C. Okoniewski. 2001. Necropsy findings and environmental contaminants in Commons Loons from New York. *Journal of Wildlife Diseases* 37: 178–84.

Windels, S. K., E. A. Beever, J. D. Paruk, A. R. Brinkman, J. E. Fox, C. C. MacNulty, D. C. Evers, and L. S. Siegel. 2013. Effects of water level management on Common Loon (*Gavia immer*) nesting success. *Journal of Wildlife Management* 77 (8): 1626–38.

12

Loon Watch

Adapting to a Changing World

I pulled up to the public boat launch. The lake water was pea green. A posted flyer warned about swimming in the lake. Don't worry, I said to myself, I have zero intentions of doing so. It was spring, and I lifted my binoculars and scanned the water, keeping an eye out for loons. Back at my car, I looked at my notes, written several months ago at my office in Maine: *Loons nested formerly on lakes in northcentral Iowa, such as Rice Lake (Winnebago County), Clear Lake (Cerro Gordo County) and Eagle Lake (Hancock County). The last reported loon pairs were on Clear Lake in 1902.* I exited the car and sat down by the lake in northern Iowa. I was working on a Common Loon restoration project with Dave Evers at the Biodiversity Research Institute (BRI). We were investigating lakes that historically had breeding loons where we might be able to introduce loons, if suitable habitat existed. I felt profoundly fortunate to be visiting these Iowa lakes, reliving history, and imagining what the water and activity were like over a hundred years ago. I recalled a passage from a Laura Ingalls Wilder book in which she recalls her Pa saying in 1850, "Minnesota is getting too crowded. Might be time to move." What would he think if he were standing beside me today?

Historically, Common Loons nested as far south as northern California, Iowa, Illinois, Indiana, Ohio, and Pennsylvania. The last confirmed nesting record for these states was forty to one hundred plus years ago. A pair nested in Connecticut in 2017 but had not bred previously there for more than fifty years. In the mid-1900s, loons nested in southern Michigan, Wisconsin, and Minnesota. The overall pattern has been a northward retraction of the breeding range over the past hundred years. Biologists have identified several factors to account for this retraction: indiscriminate shooting, an increase in human disturbance (recreational pressures), loss of suitable nest sites because of increased shoreline development, an increase in loon nest predators, and the proliferation of carp in southern Minnesota lakes.

In the decades just prior to and shortly after 1900, loons were commonly killed for sport. Hunters along the Eastern Seaboard shot them during migration because they provided fast-moving targets. Because they eat fish, loons were also seen as a threat to anglers during this time. In the late 1800s at Rice Lake, Iowa, one homeowner proudly shared information that he kept shooting adult Common Loons for several consecutive years on his lake until none came back. Unlike other birds, loons do not disperse far from their natal lakes, and once the adults were shot, there were none left to return. Indiscriminate shooting, therefore, was likely a contributory factor in the northward retraction of the loon range, despite available suitable nesting habitat. Today, it is against the law to shoot a Common Loon because they are protected by the federal Migratory Bird Treaty Act of 1918.

My first experience with breeding loons was on a wilderness lake, and I will always associate them with it. Whenever I see loons on lakes with excessive shoreline development and lots of recreational activity, I wonder how they can raise a family, let alone survive. Do recreational pressures like boating and camping affect a loon's ability to successfully hatch and fledge young? Several studies have examined this question. Minnesota's Project Loon Watch has classified the type and extent of recreational use on lakes (from no boating to heavy lake traffic, including water skiing) and the number of loon young fledged per pair (productivity). Their study found no significant difference in loon productivity between those nesting on a busy lake or one with little activity. In a separate study from 1981, conducted in the Boundary Waters Canoe Area Wilderness (BWCAW), loons did just as well on lakes with high or low canoe traffic (although this pattern broke down the following year after the study ended). The same lakes in the original BWCAW study were investigated thirty years later, and despite large increases in human disturbance between the studies, the loon population actually increased. Similar research from Michigan found that loons nesting near popular canoe routes had lower nesting success than those nesting in areas with less canoe traffic.

Can loons habituate to human activity? Mostly, yes. A study in Alaska found that loons nesting on canoe routes flushed (left the nest when

A hunter poses with a loon shot on Ragged Lake in Maine, 1895. Photograph by John W. Dunn. Collections of the Maine Historical Society.

approached by humans) at greater distances on remote lakes than loons nesting on less remote lakes, suggesting some level of selection. I have observed loon pairs nest and hatch young in plain sight of a well-used public boat launch. A nesting loon reacts to its environment, much like a robin or phoebe that nests near the back door of your garage or house. Too much disturbance upsets the rhythm of pairs switching incubation duties, especially if the disturbance takes place during the first ten days of incubation. Video surveillance cameras on loon nests in Voyageurs National Park captured people getting out of their boats and walking on islands with nesting loons. Naturally, the loons were disturbed and left the nest unattended. The study showed that if the people stayed for less than five minutes, the loons generally got back on the nest within fifty minutes, but if people stayed longer than five minutes, the loons took more than an hour to get back on. A study in Ontario found that once eggs hatched, the survival of chicks to fledging was the same for lakes that had low or high levels of boating activity. Juvenile loons on lakes with human presence showed less fear and approached closer to boats than those raised in the absence of humans.

Loons and people can coexist; however, a loon pair cannot raise a family if they are being perpetually disturbed. There is a breaking point. Recreational boating is a greater threat to loons in open water than to those nesting in coves and foraging in shallow water. Studies showed 39 percent of all loon mortality in New England was caused by trauma due to boats. Many loons, especially those with young, cannot get out of the way of high-speed watercraft. Loons can habituate to recreational activity, but the degree of habituation likely varies by the tolerance levels of individuals, pairs, loon territory, type of lake, type and amount of disturbance, prey abundance, and a suite of other factors. The data suggest that increased human recreational pressure has led to some range retraction in recent decades, but I think that loss of shoreline habitat may have been the single most important contributing factor.

Shoreline development alters lake habitat, which directly affects resident loons. Vegetated shoreline habitat is critical to attracting fish. Removing nearshore vegetation and replacing it with lawns leads to lake eutrophication. Eutrophication is a process where a lake receives excessive nutrients (phosphorus and nitrogen) from fertilizers and agricultural runoff, which leads to an increase in plant growth, primarily microscopic algae. These algae grow quickly and form a dense mat on the lake. The algal mat blocks sunlight from entering the water, limiting or preventing other submerged aquatic vegetation from photosynthesizing, eventually killing them. These plants live beneath the surface of the water, much like a kelp forest in the ocean, and they provide critical habitat for many fish species. Submerged aquatic vegetation comes in many forms and constitutes a taxonomically diverse assemblage of plants, such as coontails, milfoils, and grasses. Algae themselves are short-lived, and they eventually die and settle at the bottom of the lake, where bacteria and fungi decompose them. This decomposition process consumes oxygen, and over time, the lake will contain less of the dissolved oxygen needed to support submerged aquatic vegetation and native fish populations. Coldwater fish, such as trout, require rich oxygenated water, while panfish (sunfish, pumpkinseed) and carp can survive better in water with lower oxygen. Anne Kuhn, a researcher with the Environmental Protection Agency, identified negative associations between the use of lakes by Common Loons and the presence of agricultural and developed lands, lake total phosphorus concentration, road density, and proximity to human population centers.

Boat ramps specifically influence the type and amount of submerged aquatic vegetation, directly affecting lake fisheries. For example, they allow for transference of non-native invasive aquatic plants, such as Eurasian milfoil, from one lake to another. Non-native milfoil grows densely and outcompetes other submerged aquatic vegetation, thereby reducing biodiversity and influencing local fish populations. Many fish species depend on submerged aquatic vegetation for their survival because the vegetation provides critical nursery habitat. Submerged vegetation also attracts an abundance of invertebrates, prey for many juvenile game and nongame fish species. These, in turn, attract larger predatory fish, such as yellow perch, which are preyed on by loons. Milfoil also consumes large quantities of dissolved oxygen at night, thus lowering oxygen levels. Native fish species, such as trout and perch, do not fare well under these conditions, but carp and bullheads do. Carp and bullheads root in silty lake bottoms and stir up nutrient-laden sediment. Murky water blocks sunlight from reaching aquatic plants that stabilize the lake bottom and provide oxygen and habitat for game fish and aquatic insects. Lakes and reservoirs dominated by carp and bullhead have little submerged aquatic vegetation and low visibility, whereas lakes with submerged aquatic vegetation have higher water clarity and greater fish diversity.

As the amount of lakeshore development and human recreational pressure increased, loons left historical nesting and nursery locations for less suitable ones, resulting in fewer chicks fledged per territorial pair. Researchers in the Adirondack Mountains of New York have found that development on smaller lakes had a more significant effect on loon nesting success than development on larger lakes. Other studies showed that development alone (e.g., homes), without the concomitant human traffic and activity, had little effect on loon reproductive success. However, development *with* an increase in human traffic typically led to lower nesting success and productivity. The mixed results are likely due to varying levels of habituation in loons, both spatially and temporally. Over generations, selection has favored loons that can tolerate and habituate to human disturbance over individuals that cannot adapt or were slower to habituate. When I talk to longtime lakeside residents in Maine and New York, I hear the same story: the loons have habituated to people and their watercraft. "We could never get this close to them twenty or thirty years ago," they say. Recreational activity likely had an initial negative impact on loon productivity in many parts of the country, but over time, as a new generation of loons habituated to activities like boating and camping, productivity returned to previous levels.

However, shoreline development often leads to an increase in loon egg predators such as raccoons and gulls. Some studies have found that raccoons predated 50 percent of the loon nests in a local area. A high rate of predation may occur on lakes with just one or two spots where loons can successfully nest, and predators naturally target such sites. The loss of suitable habitat, coupled with an increase in egg predators, has most likely played a significant role in the disappearance of nesting loons from the southern part of their range over the past one hundred years.

Reservoirs and Rafts

•

The construction of dams has also affected the loon landscape. By holding back water to generate electricity, dams alter a lake's natural fluctuating water levels and its rate of change. For example, instead of a big spring runoff, dams hold back the water and release it slowly over the summer. In these cases, loons initiate nesting during the peak of high water only to have water levels drop down unnaturally low, leaving the nest stranded. Sometimes, the opposite happens. For example, in some years dam operators cannot retain enough water and have to release a large amount, resulting in a drastic rise in water levels. Loons can tolerate some rising water levels by building up the lip of their cup nest but only to a point; anything above 6 inches in three to four days is too much. Water levels increasing 6 inches (or greater) or decreasing 12 inches (or greater) within thirty days after a pair initiates a nest will negatively affect nesting success. These fluctuations likely affect the prey base too. Quick drops of water levels can kill the eggs and larvae of fish that spawn in shallow littoral habitats. It is not surprising, then, that reservoirs in Minnesota, New York, and Maine had the lowest reproductive success rates (number of chicks hatched/territorial pair) of any waterbodies in the nation.

In 1975, Rawson Wood, a retired businessperson living on Squam Lake in New Hampshire, was concerned about the decline of loons in the state. He wrote a proposal to the New Hampshire Audubon Society to establish a committee that would monitor and protect them. The society approved the proposal and appointed Rawson chair of the committee; they called themselves the Loon Preservation Committee (LPC). In 1978, LPC hired Jeff Fair as their first biologist to monitor and look after the loons on Lake Umbagog. Shortly after the committee's formation, staff and board members discussed using floating artificial islands to improve loon nesting success on reservoirs. By creating a platform of logs and filling the spaces with moss, grasses (sedges), and other vegetation, they created artificial islands on which loons were able to nest. These floating island rafts were protected from fluctuating water levels and mammalian predators, and breeding success on New Hampshire reservoirs soared. Because their deployment on reservoirs was so successful, rafts were also used on stable waterbodies with high shoreline development or those with a lack of islands. Today, rafts are a common conservation tool used by agencies, companies, and private citizens across the nation. However, in some ways, rafts can do more harm than good. The deployment of too many rafts and in the wrong locations can lead to increased male–male contests and lower overall reproductive success.

Knowledge of previous traditional nest sites and territorial boundaries is essential for determining how many rafts to put out and where they should go. Rafts should be placed strategically to minimize their exposure to wind and waves, boat traffic, and patterns of human activity. When biologists noticed that incubating loons on rafts are more vulnerable to eagle predation, and their eggs more vulnerable to crow and gull predation, overhead guards were added to minimize these threats. They consist of mesh or camouflage netting strung over the nest and have proven successful in reducing exposure and avian predation. They also allow loons to feel more protected and to flush less often when recreationists come too close. In high-use areas, signs, buoys, and roping can be effective in reducing or eliminating activity around the platforms. To maximize public compliance, ropes are best removed immediately following hatching, which requires constant monitoring.

Adaptive Water Level Management

•

The first time I banded loons in Voyageurs National Park was August 1993. The park is located on the border of Minnesota and Canada and is named in honor of the French Canadian fur trappers who navigated these waters in the 1800s in their large canoes. Voyageurs National Park is atypical because it is mostly water and its lakes occupy both the United States and Canada. My team and I were out at night, having little success at loon catching. We were unfamiliar with the loon territories on this big waterbody and thought that a faint yodel or tremolo might indicate where some of the pairs were located. We shut off the engine and listened for them. We looked up at the sky, figuring that we would spot the familiar dominant stars (Polaris, Antares, Vega, Altair) but instead got the surprise of the summer—the northern lights! This was just my second time seeing them, and I was intrigued by their color and movement—shades of green, shaped like a serpent, moving mysteriously across the sky. Twenty-six years later it remains one of my most memorable moments while catching loons.

Before the park was created in 1975, the area was operated by Minnesota Power, which in the early 1900s built a series of dams to generate hydroelectric energy. Over time those dams changed the shape of some of the existing lakes, including Rainy, Namakan, Kabetogama, and Sand Point. Because these lakes are border waters, an International Joint Commission (IJC) was created to regulate their water levels through management programs known as rule curve. The 1970 rule specified maximum and minimum water levels as well as the timing of peak water level. For example, water levels on Namakan Reservoir would peak in late June or early July and be allowed to fluctuate roughly 10 feet (3 meters), but on Rainy Lake they were allowed to fluctuate only 3.3 feet (1 meter). Under the estimated pre-dam conditions, the fluctuation range on both lakes was likely 3 to 4 feet, with water level peaks on Namakan between mid-May and early June, and on Rainy between late June to early July. What did this mean for loons in the area?

A 1988 study by the National Park Service found that reproductive success of Common Loons on Rainy Lake was good (0.57 fledged young/territorial pair) but poor on Namakan Reservoir (0.25 fledged young/territorial pair). The greatest cause of nest loss was flooding. Because of this research, in January 2000 the IJC adopted a revised water management program that incorporated a more natural hydrologic regime. The 2000 rule specified a peak water level on Namakan Reservoir by May 31 (moved up one month) with fluctuations of approximately 2 meters (instead of 3; approximately 6 versus 9 feet). No changes were made on Rainy Lake. Under the new 2000 rule the loons on Namakan fared much better. Overall, loon productivity (chicks hatched/territorial pair) increased 95 percent on Namakan between the two time periods, 1983 to 1986 and 2004 to 2006. It appears that the factor limiting loon nest success was the timing of peak water level. If the rule was followed and water levels peaked by May 31, then loon productivity increased. This example illustrates how active park management, without deploying platforms, can have a positive effect on loon productivity.

These findings tell us that not all lakes nor all situations are equal. I support and applaud grassroots organizations taking the initiative to deploy rafts and monitor the loons on their lakes, but this activity must be supervised to avoid adverse situations that lower loon reproductive success. Platforms help loons best when citizens groups work with knowledgeable and established loon conservationists or state officials to plan for the number of rafts and where to deploy them. This will vary from lake to lake.

Introductions and Restoration

•

Animal reintroductions in the United States have been taking place for at least one hundred years with the goal of returning native species to their former habitats if they were extirpated by overexploitation or habitat degradation. The term *reintroduction* is well established in conservation biology, but it is perhaps misleading since it implies that an animal has to be introduced in the first place in order to be reintroduced a second time. Avian reintroductions have involved both game (Ruffed Grouse, Wild Turkey, Prairie Chicken, Sandhill Crane) and nongame species (Eastern Bluebird, Trumpeter Swan, Peregrine Falcon). Many of these reintroductions have been successful, as defined by creating a self-sustaining population after a period of three (or more) years. In the simplest case, one catches young or adult individuals from the source population and releases them in the experimental zone, which may be several hundred miles away; but in complex cases, reintroduction may entail actually feeding and raising young in captivity and teaching them the migration route using light aircraft (e.g., Sandhill Cranes). Several factors have been identified as critical to the success of reintroductions. These include suitable habitat availability, awareness of the behavior of the animal in its new environment (e.g., potential stressors, like boating), a large initial founder population to provide genetic diversity, and support by relevant wildlife stakeholders such as state and federal agencies and those in the public and private sectors.

Reintroductions had never been tried with loons, and a number of people were interested to see if it could be done, primarily as a conservation management tool. Common Loons appeared to be ideal candidates for reintroduction efforts, and in 2013 Dave Evers wrote a successful proposal to obtain funding for such a project. The best place in the contiguous United States to attempt this initially was Minnesota, for a couple of reasons. First, there would be no permitting issues; Minnesota has more loons than all the other states in the Lower 48 combined (more than ten thousand). Second, breeding loons have been extirpated from the southern part of the state. Third, there appeared to be several lakes in the extirpated range that could support nesting loons. Fourth, the state has a highly supportive Nongame Wildlife Program in the Department of Natural Resources (DNR).

In early 2013, Dave and I flew to Minnesota to meet with the program's supervisor, Carrol Henderson, and other key players in the DNR to discuss the reintroduction plan. After a successful meeting and approval to proceed, I worked with analyst Jeff Tash at BRI to generate a list of possible lakes for reintroduction based on several criteria: we needed a cluster of lakes with low to moderate shoreline development, several suitable nesting locations, fairly low human recreational pressure, a healthy native fish population, and low exotic plant population. We sent this list to the DNR, and with their knowledge and expertise they made valuable suggestions. That summer, field biologist Nick Elger and I visited and evaluated 307 lakes in twenty-seven counties as possible translocation sites. In the end, we chose a small cluster of lakes (e.g., Fish, Roemhildts) in Le Sueur County, west of Faribault in southern Minnesota. We identified remote lakes approximately 240 miles to the north, in Itasca County, to be the source population for the chicks. Beginning in the fall of 2014 and throughout the winter, our team worked in partnership with the DNR on the reintroduction plan.

Michelle Kneeland, BRI wildlife veterinarian, led discussions on what type of enclosure to use that would keep the loon chicks protected from predators yet be sturdy enough to withstand wave action. In the end, we obtained two heavy galvanized steel wire mesh cages measuring 8 feet tall, 12 feet wide, and 24 feet long, each weighing more than

800 pounds. We lined the part of the cage below the water level with fine mesh to keep fish in and underwater predators out. We selected locations for the pens based on availability of a shoreline vantage point where we could monitor the loons, a water depth of 3 to 5 feet, and a lack of aquatic macrophytes to give the chicks space for underwater swimming. The nearest access point was 150 yards away: How were we going to get the pens in place? We called Joe and Tony Musial, owners of a local underwater salvaging company, for assistance. After a lot of deliberation and head scratching, Tony came up with a brilliant suggestion: attach barrels to the pens so they could float. Then we could pull the pens by boat and drop them in the water at the specified location. This worked, but the process was labor intensive. The pens were placed 10 to 15 feet apart, with the doors facing away from shore. We connected a 75-pound anchor to each corner to ensure the pens remained in place, and we covered the top with avian guard netting to keep predators out. The area immediately surrounding the pens was roped off from the rest of the lake with a regulatory buoy that cautioned boaters and any curious onlookers to avoid the area.

We all agreed the translocated chicks should be between six and ten weeks of age. Chicks any younger would still require a fair amount of parental care, and older chicks can be very hard to catch. Theoretically, a younger chick would also spend more time than an older one on the release lake, thereby increasing the odds of imprinting on it. Imprinting in this case is a process where recently hatched young come to think of a certain place as home, developing a visual or olfactory memory of where home is and what it looks like. After all, the goal was to get the loons to return to the release lake, not the natal one. The night of August 14 and morning of August 15 were memorable: it was the first time in history that a loon chick was transported for relocation. All the BRI staff worked long hours, but Michelle in particular was indefatigable, working twenty-one consecutive hours. To feed the chicks, we relied on several local bait shops. On August 29, we released the first chick from the pen on Fish Lake. It had been in captivity for fifteen days and flew off the lake on September 4. We thought it would remain on the water longer than it did, and openly wondered if it had spent enough time on the lake to imprint on it.

We translocated five loon chicks during summer 2014. Needing to track them but not wanting to cause them additional stress, we opted for a small leg-attached radio transmitter instead of a bulky surgically implanted satellite transmitter. We were able to pick up a signal from one of the translocated chicks at a larger, adjacent lake, but the majority were never detected again once they flew off the release lake. Our team checked for them on their natal lakes but never spotted them. All we could do was wait. The following winter we decided to adopt a hybrid approach of raising some chicks in pens while releasing others directly onto the lake. In 2015, we released four pen-raised chicks and three chicks directly. The following year we opted to do direct release only and transported five additional chicks. For the three-year period, a total of seventeen loons were translocated successfully. Did any loons come back? In 2017, a banded loon was observed on Fish Lake for a couple of weeks in June, but BRI was unable to confirm whether it was one of our translocated birds (though it certainly looked promising).

There was a second phase to the project, which took place in Massachusetts. Over three years (2015–17), thirteen loon chicks were moved from New York and Maine, and at least four of them were confirmed to have survived to adulthood and returned to the release lake. All the hard work paid off: the first ever documented case of a translocated chick reaching adulthood, breeding, and producing a chick was documented in 2020. In 2015, a male chick, 6.5 weeks old, from upstate New York, was driven approximately 203 miles to Pocksha Pond in Middleborough, Massachusetts, where it was reared for twenty-three days in a pen before being released. It was observed on Pocksha Pond in both 2018 and 2019 and paired with an unbanded female in 2020 on Copicut Reservoir (Fall River, Massachusetts), approximately 11 miles south of its release lake, where the pair hatched a single chick.

Global Climate Change

•

Several climate models predict that in the next forty years most of the current Common Loon breeding range will become warmer and in some parts drier. Warmer ambient temperatures will lead to higher lake temperatures, resulting in eutrophication and a concomitant shift in lakes' fish community structure. These changes do not bode well for loons. Just like other birds, loons require enough food to provide for their chicks as well as themselves in order to sustain their population.

Beyond that, they need suitable habitat for nesting and brooding. For loons, climate change is unlikely to impact nesting sites, especially if platforms are available, but warmer temperatures may cause them to get off the nest more frequently and may disrupt normal incubation patterns (see chapter 5). This has the potential to lower loon nesting success. Water clarity will also be affected, and this too may impact a loon's ability to find food efficiently by influencing the type and amount of food it can provide for its young, which will likely lower offspring survival. Worse, if foraging becomes too difficult, adult loons may not be capable of providing enough food for themselves, let alone their young.

Audubon scientists have developed a model using more than one hundred environmental and habitat data points to predict how climate change may affect the breeding range of several bird species, including loons. The model predicts that both the summer and winter range of loons will recede northward and that Common Loons could lose as much as 56 percent and 75 percent of their current summer and winter ranges, respectively, by 2080. The model goes so far as to suggest the possibility of loons disappearing completely from Minnesota by the end of the twenty-first century. I applaud Audubon scientists for their forward thinking and for taking a stab at a difficult task, but I wonder if the model took into account that some loons have expanded their ranges to the south, to areas where they were formerly extirpated, such as Massachusetts, Connecticut, and central Wisconsin. I acknowledge there is no clear predictable outcome and that none of us have a good handle on the rate of environmental changes that will occur in the next decades. That, perhaps, is the most frightening realization.

Although loons are primarily viewed as occupying and preferring large oligotrophic lakes, they do nest and occasionally raise a family on small eutrophic lakes. Researchers in Alberta found this to be the case, even observing loons raising young on fishless lakes. They noted that loons nesting on larger lakes with fish ultimately fared better than those nesting on smaller fishless lakes. In 2013, when I was doing loon surveys across southern Minnesota, I was surprised to find loons present and even breeding on lakes that were much more eutrophic than lakes where I am accustomed to seeing loons in the upper Midwest and New England. This gives me hope that loons may be able to adapt to a level of increasing eutrophication, but I seriously question how sustainable that population will be in the long term.

Finally, as the planet warms, there is the real possibility of tropical diseases spreading northward. For example, Mark Pokras, an emeritus professor at Tufts University, and Ellen Martinson, a researcher with the Smithsonian Conservation Biology Institute, found malaria parasites in New England loons ten to fifteen years ago, and in 2014 a loon died from malaria. This occurred on Umbagog Lake on the border of New Hampshire and Maine. The concern, naturally, is how much resistance loons have to malaria, a disease with a long and devastating history. In 1826, while the ship *Wellington* was docked at Maui, deckhands drained water from barrels that likely carried the night-flying mosquito, a vector of the malaria parasite *Plasmodium.* Prior to this time, the island lacked mosquitoes. Because birds migrate, some were likely already carrying the malaria parasite, but it needed a vector, such as mosquitoes, to transport it. Having little resistance to the disease, many of the island's native birds likely died from malaria, as has been suggested by reports of dead birds littering the forest floor. It is unclear how loons exposed to malaria will respond, but the threat certainly raises eyebrows. Mark Pokras and other researchers are paying close attention to loon mortality.

In summer 2019, an Earthwatch volunteer, Sherry Abst, kayaked out to a loon that was acting strangely in northern Minnesota, outside Ely. It later died of West Nile virus. Two more loons were collected, and they too died from West Nile. Arno Wuenschmann, a veterinarian with the University of Minnesota, noted that these were not the first cases of West Nile reported in the state. He had documented an entire family of four loons that died from West Nile virus fourteen years earlier on Sandy Lake near Zimmerman. West Nile virus, originally documented in Africa, first showed up in Queens, New York, in 1999 and has spread from there. Certain groups of birds, such as jays, crows, and northern owls, are particularly susceptible to it. The virus cannot be treated and kills quickly, in two to three weeks. If you find a loon acting sickly on a lake, contact your local DNR office. This is another disease that has loon aficionados on high alert.

Can we tell if the effects of climate change, such as these emerging diseases, are causing population-level effects in loons? Are just a few individuals in the population dying, or is the problem more widespread? One long-term study analyzing loon data gathered from 1,500 lakes over thirty-eight years in Ontario found a decline in loon productivity (number of chicks fledged per pair) over that time span. The study concluded that the decline was largely due to low pH and higher than normal mercury levels, both of which may be exacerbated by climate change, but that climate change alone was not the main driver of the changes. If climate change were responsible, one would predict that regions in the south would undergo population changes before those in the north.

Fortunately, we can put systems in place to detect if such a change has occurred. An example is the Minnesota Loon Monitoring Program, begun in 1994, when Eric Hanson, then a master's student at the University of Minnesota, worked with the Minnesota DNR to developed a long-term plan to monitor the statewide loon population. Elegantly designed, the plan divided the state's loon population into six zones, randomly selected one hundred lakes from each zone, and, using citizen scientists, monitored those same lakes at the same time each year for adults and chicks. The results of this intensive effort do not show any major shifts or changes in the loon population in any of the zones from 1994 to 2015, which is encouraging. If, however, the population does decline, officials in Minnesota will be the first to know.

During his day, Charles Darwin was unaware of genetics, but he knew that some hereditary factor was involved in the shape, behavior, and life histories of organisms. Today, we know that the ability of a population to remain healthy and adapt to change depends on its genetic diversity (see chapter 4); that is, the higher the genetic diversity among individuals in the population, the more it will be able to cope with a parasite or virus. Typically, populations that range over continents, such as loons, have a high degree of genetic diversity, while populations inhabiting islands have low genetic diversity. Amy McMillan, at Buffalo State University, took the first stab at determining the genetic diversity of Common Loons and concluded that they appear to have low genetic diversity. A later study by Alec Lindsay at Northern Michigan University, using newer techniques and a larger data set, also concluded that loons have low genetic diversity compared to other species that range across the continent, such as Mourning Doves and chickadees.

Why might loons have low genetic diversity? Over the millennia, glaciers covered North America several times, and each time they descended southward they extinguished suitable habitat (breeding lakes). Loons were forced to move southward to find waterbodies on which to nest and reproduce. Fossils from birds with a northern range similar to loons', such as Spruce and Ruffed Grouse, show up in southern states, such as Georgia and Florida. There are few natural lakes in the southeast. Lakes were predominately formed when glaciers scoured out the bedrock, and since the glaciers did not reach that far south, there were likely few lakes for loons to occupy. I imagine that competition for what existed at the time was intense and only a small subset of the much larger loon population survived. When the glaciers retreated northward, the small population of loons that survived the glacial episodes began to expand their range northward, and with new lakes available for breeding, the population increased, eventually occupying lakes as far north as Canada. I suspect that the low genetic diversity of loons is the result of a historical constraint. Will this low genetic diversity impact how loons cope with malaria and other emerging diseases? Only time will tell.

Where do we go from here? For several reasons, we can expect that loons and people will continue to coexist in the distant future. First, on some of the busiest lakes in New England, such as Squam Lake in New Hampshire and Little Sebago Lake in Maine, loons still produce chicks annually, despite exceedingly heavy boat traffic and extensive shoreline development. Second, loon populations are growing and expanding in some states. They are nesting once again in central Wisconsin, southern Massachusetts, and northern Connecticut, where they have been absent for more than fifty years. Third, our knowledge, understanding, and public awareness of loons is at an all-time high. There are more loon conservation organizations, conferences, and festivals than there have ever been. Loon organizations work with state officials to monitor loon populations. Loon conferences provide a forum and opportunity for scientists to share what they have learned, help direct future research and management objectives, and identify potential environmental threats. Loon festivals foster appreciation for these magnificent birds and may excite or inspire the next generation of conservationists. I am optimistic that loons and people will continue to coexist into the next century mostly because loons affect people in a way that most birds do not. People love their loons and will go to great lengths to make sure they grace our lakes each year.

Several people have inspired me in their devotion and commitment to loon conservation and education. I have known Dan Poleschook and Ginger Gumm since 1993, when I was in Washington State to sample mercury in loons. Dan and Ginger are professional photographers who have spent an inordinate amount of time on the water getting to know loons and their habits. In short order these loon enthusiasts became loon experts. They noticed that some adult loons were dying from lead ingestion (see chapter 11) and notified the Washington Department of Fish and Wildlife. Over the years they have given outreach talks throughout the state and continue to petition officials to take action. In 2009, Dan and Ginger presented their data on loon mortality due to lead tackle and other fishing activities during public hearings to the state wildlife commissioner. They faced outright hostility from fishing lobby groups. In 2010, the Washington Department of Fish and Wildlife named Dan and Ginger Educators of the Year for their leadership and role in educating the public and the state about lead mortality in loons. That same year the department passed a lead tackle ban, excluding its use during the loon breeding season on twelve of the thirteen lakes where they nest annually.

I met Brooks Wade the winter of 2014, when I visited him at Lake Jocassee, a South Carolina reservoir in the heart of the Appalachian Mountains. Here, Brooks leads boat tours, educating visitors and residents about the natural history and wonders of the Lake Jocassee region. National Geographic lists the Jocassee Gorges as one of the fifty great places on Earth. What brought me to visit Brooks was the opportunity to do research on the lake's wintering loon population. He is a native Floridian and has spent much of his adult life in the fishing and restaurant industries, but he always loved watching winter loons off the coast. His wife, Kay, vacationed at Lake Jocassee and fell in love with it. They decided to move there and before long started a small boat tour business. Over the past ten years, he and his staff at Jocassee Lake Tours have educated thousands of visitors about its wintering loons. His weekly newsletter reaches more than a thousand people. Brooks has asked me more questions about loons than the next twenty people combined. His knowledge and hunger for learning are inspirational. His enthusiasm for life is contagious. I wrote this book for people like him.

In 1993, Ari, a high school student from Texas, was my assistant at Seney National Wildlife Refuge. In late May we were checking on a loon nest on an island and noticed that the eggs were outside the nest next to the water. Ari wanted to swim to the island, probably 30 to 40 feet away, and put them back. He swam out, crawled up the bank, and moved the eggs back into the nest bowl. "Jim, I'm so cold!" he yelled to me. He was less than halfway back when he shouted, "Jim, Jim! I'm drowning! Help me!" I dove in, and to Ari's credit, he did not panic and let me support his head and chest out of the water. When we reached the shore, I could see he was shaking. With the car heater on full throttle, I drove back to the cabin. Ari worked hard all the rest of that summer. What happened to the loon eggs? Both hatched successfully.

I tell this story about Ari because he showed empathy, one of our greatest virtues. Why do humans feel compelled to help loons, wildlife, or other people? Because at our core we are compassionate. Although compassion is written into our DNA, it needs to be modeled and cultivated for it to take root. Loons remind us with their constant vocalizing and stunning appearance that we share this planet with other inhabitants. Their presence fosters awareness of others, the first step to developing a more holocentric worldview, in which decisions are based on what is best for the community. Such a worldview exists in stark contrast to an anthropocentric worldview, which prioritizes what is best for the individual. Anthropocentrism is often the root cause of ecological and planetary problems created by human action. Without environmental and experiential learning, some of us will think, "It is all about me." But with the proper exposure to nature, maybe some of those people will become passionate stewards who look after our lakes and our loons.

If you love these birds as much as I do, I encourage you to work to protect them. Many states with breeding loons have volunteer loon monitoring programs that are doing great work, such as the Minnesota DNR's Loon Monitoring Program, Wisconsin's LoonWatch through the Sigurd Olson Environmental Institute, the Maine Audubon's Loon Project, the Vermont LoonWatch program, and the Montana Loon Society. Organizations like these can always use volunteers to count adult loons and chicks. Lake associations and state agencies also collect data on water quality and can use volunteers. Clean lakes are vital to maintaining a robust loon population. Important conservation work is also being done by the Loon Preservation Committee, the Adirondack Center for Loon Conservation, the Minnesota Pollution Control Agency, and the

Biodiversity Research Institute. These organizations can use your support to conserve loons for future generations. Collectively, every little bit helps.

To Dan, Ginger, Brooks, and Ari: we got this.

Further Reading

Badzinski, S. S., and S. T. A. Timmermans. 2006. Factors influencing productivity of Common Loons (*Gavia immer*) breeding on circumneutral lakes in Nova Scotia, Canada. *Hydrobiologia* 567: 215–26.

Baker, L. A., J. E. Schussler, and S. A. Snyder. 2008. Drivers of change for lake-water clarity. *Lake and Reservoir Management* 24: 30–40.

Barr, J. F. 1996. Aspects of Common Loon (*Gavia immer*) feeding biology on its breeding ground. *Hydrobiologia* 321: 119–44.

Bent, A. C. 1919. *Life Histories of North American Diving Birds.* Smithsonian Institution, U.S. National Museum Bulletin 107, U.S. Government Printing Office, Washington, D.C.

Bianchini, K., D. C. Tozer, R. Alvo, S. P. Bhavsar, and M. L. Mallory. 2020. Drivers of declines in Common Loon (*Gavia immer*) productivity in Ontario, Canada. *Science of the Total Environment* 738: 139724.

Blair, R. B. 1992. Lake features, water quality and the summer distribution of Common Loons in New Hampshire. *Journal of Field Ornithology* 63: 1–9.

Carlson, R. E. 1977. A trophic state index for lakes. *Limnology and Oceanography* 22: 361–69.

Church, J. A., and N. J. White. 2006. A 20th century acceleration in global sea level rise. *Geophysical Research Letters* 33, L01602, doi:10.1029/2005GL024826.

Dahmer, P. A. 1986. Use of aerial photographs to predict lake selection and reproductive success of Common Loons in Michigan. Master's thesis, University of Michigan, Ann Arbor.

Deguchi, T., J. Jacobs, T. Harada, L. Perriman, Y. Watanabe, F. Sato, N. Nakamura, K. Ozaki, and G. R. Balogh. 2012. Translocation and hand-rearing techniques for establishing a colony of threatened albatross. *Bird Conservation International* 11: 66–81.

DeSorbo, C. R., J. Fair, K. Taylor, W. Hanson D. C. Evers, and H. S. Vogel. 2008. Guidelines for constructing and deploying Common Loon nesting rafts. *Northeastern Naturalist* 15 (1): 75–86.

DeSorbo, C. R., K. M. Taylor, J. Fair, D. Kramar, J. L. Atwood, D. C. Evers, W. Hanson, and H. S. Vogel. 2007. Reproductive advantages for Common Loons (*Gavia immer*) using rafts. *Journal of Wildlife Management* 71 (4): 1206–13.

Dinsmore, J. J., T. H. Kent, D. Koenig, P. C. Petersen, and D. M. Roosa. 1984. *Iowa Birds.* Ames: Iowa State University Press.
Forbush, E. E. 1927. *Birds of Massachusetts and Other New England States.* Massachusetts Department of Agriculture.
Found, C., S. M. Webb, and M. S. Boyce. 2008. Selection of lake habitats by waterbirds in the boreal transition zone of northeastern Alberta. *Canadian Journal of Zoology* 86: 277–85.
Gingras, B. A., and C. A. Pazkowski. 1999. Breeding patterns of Common Loons on lakes with three different fish assemblages in north-central Alberta. *Canadian Journal of Zoology* 77: 600–609.
Grinnell, J., and A. H. Miller. 1944. *The Distribution of the Birds of California.* Pacific Coast Avifauna, No. 27. Berkeley, Calif.: Cooper Ornithological Club.
Hammond, C. A., M. S. Mitchell, and G. S. Biselle. 2012. Territory occupancy by Common Loons in response to disturbance, habitat and intraspecific relationships. *Journal of Wildlife Management* 76 (3): 645–51.
Harlow, R. C. 1908. Breeding of the loon in Pennsylvania. *The Auk* 25: 471.
Heimberger, M., D. Euler, and J. Barr. 1983. The impact of cottage development on Common Loon reproductive success in central Ontario. *Wilson Bulletin* 95 (3): 431–39.
Hoffman, R. 1910. *A Guide to Birds of New England and Eastern New York.* Cambridge, Mass.: Houghton-Mifflin.
Jung, R. E. 1987. An assessment of human impact on the behavior and breeding success of the Common Loon (*Gavia immer*) in the northern Lower and eastern Upper Peninsulas of Michigan. Master's thesis, University of Michigan, Ann Arbor.
Kaplan, J. D. 2003. Human recreation and loon productivity in a protected area, Isle Royale National Park. Master's thesis, Michigan Technological University, Houghton.
Keast, A. 1984. The introduced aquatic macrophyte *Myriophylum spicatum,* as habitat for fish and their invertebrate prey. *Canadian Journal of Zoology* 62: 1289–1303.
Kelly, L. M. 1992. The effects of human disturbance on Common Loon productivity in northwestern Montana. Master's thesis, Montana State University, Bozeman.
Kenow, K., P. Garrison, T. Fox, and M. W. Meyer. 2013. Historic distribution of Common Loons in Wisconsin in relation to changes in lake characteristics and surrounding land use. *Passenger Pigeon* 75: 375–89.
Kuhn, A., J. Copeland, J. Colley, H. Vogel, K. Taylor, D. Nacci, and P. August. 2011. Modeling habitat associations for the Common Loon (*Gavia immer*) at multiple scales in northeastern North America. *Avian Conservation and Ecology* 6: 4.
Kumlien, L., and N. Hollister. 1951. *The Birds of Wisconsin.* Madison: Wisconsin Society for Ornithology.

Mager, J. N., C. Walcott, and W. H. Piper. 2008. Nest platforms increase aggressive behavior in Common Loons. *Naturwissenschaften* 95 (2): 141–47.

McIntyre, J. W. 1975. Biology and behavior of the Common Loon (*Gavia immer*) with reference to its adaptability in a man-altered environment. Ph.D. diss., University of Minnesota, Minneapolis.

McIntyre, J. W. 1979. Status of Common Loons in New York from a historical perspective. In *Proceedings of the North American Conference on Common Loon Research and Management,* edited by S. A., Sutcliffe, vol. 2, 117–21. Washington, D.C.: National Audubon Society.

McIntyre, J. W. 1988. *The Common Loon: Spirit of Northern Lakes.* Minneapolis: University of Minnesota Press.

Miconi, R. M., M. Pokras, and K. Taylor. 2000. Mortality in breeding Common Loons: How significant is trauma? In *Loons: Old History and New Findings. Proceedings of a Symposium from the 1997 Meeting,* edited by J. W. McIntyre and D. C. Evers. Holderness, N.H.: American Ornithologists' Union, North American Loon Fund.

Olson, S. T., and W. H. Marshall. 1952. *The Common Loon in Minnesota.* Minneapolis: Minnesota Museum of Natural History, University of Minnesota.

Paruk, J. D., D. C. Evers, J. W. McIntyre, J. F. Barr, J. N. Mager, and W. H. Piper. 2021. Common Loon (*Gavia immer*), version 2.0, *The Birds of North America,* edited by P. G. Rodewald. Ithaca, N.Y.: Cornell Lab of Ornithology. https://doi.org/10.2173/bna.

Paruk, J. D., J. N. Mager III, and D. C. Evers. 2014. An overview of loon research and conservation in North America. *Waterbirds* 37: 1–5.

Piper, W. H., M. W. Meyer, M. Klich, K. B. Tischler, and A. Dolsen. 2002. Floating platforms increase reproductive success of Common Loons. *Biological Conservation* 104 (2): 199–203.

Radomski, P. J., K. Carlson, and K. Woizeschke. 2014. Common Loon (*Gavia immer*) nesting habitat models for north-central Minnesota lakes. *Waterbirds* 37: 102–17.

Ream, C. H. 1976. Loon productivity, human disturbance and pesticide residues in northern Minnesota. *Wilson Bulletin* 88: 427–32.

Reiser, M. H. 1988. *Effects of Regulated Lake Levels on the Reproductive Success, Distribution and Abundance of the Aquatic Bird Community in Voyageurs National Park, Minnesota.* Reservoir Resource Management Report. MWR-13. National Park Service. Omaha, Neb.

Roberts, T. S. 1932. *The Birds of Minnesota.* Minneapolis: University of Minnesota Press.

Ruggles, A. K. 1994. Habitat selection by loons in southcentral Alaska. *Hydrobiologia* 279/280: 421–30.

Spilman, C. A., N. Schoch, W. F. Porter, and M. J. Glennon. 2014. The effects of lakeshore development on Common Loon (*Gavia immer*) productivity in the Adirondack Park, New York, USA. *Waterbirds* 37: 94–101.

Strong, P. I. V., and R. Baker. 1991. An estimate of Minnesota's summer population of adult Common Loons. Biological Report No. 37. Minnesota Department of Natural Resources.

Strong, P. I. V., and J. A. Bissonette. 1989. Feeding and chick-rearing areas of Common Loons. *Journal of Wildlife Management* 53: 72–76.

Strong, P. I. V., J. A. Bissonette, and J. S. Fair. 1987. Reuse of nesting and nursery areas by Common Loons. *Journal of Wildlife Management* 51: 123–27.

Sutcliffe, S. A. 1980. Aspects of the nesting ecology of Common Loons in New Hampshire. Master's thesis. University of New Hampshire, Durham.

Titus, J., and L. Van Druff. 1981. Response of the Common Loon to recreational pressure in the Boundary Waters Canoe Area, northeastern Minnesota. *Wildlife Monograph* no. 79. The Wildlife Society.

Valley, P. J. 1987. Common Loon productivity and nesting requirements on the Whitefish chain of lakes in north-central Minnesota. *The Loon* 59: 3–10.

Vermeer, K. 1973. Some aspects of the nesting requirements of Common Loons in Alberta. *Wilson Bulletin* 85 (4): 429–35.

Warner, R. E. 1968. The role of introduced disease and the extinction of the endemic Hawaiian avifauna. *The Condor* 70: 101–20.

Windels, S. K, E. A. Beever, J. D. Paruk, A. R. Brinkman, J. E. Fox, C. C. MacNulty, D. C. Evers, L. S. Siegel, and D. C. Osborne. 2013. Effects of water-level management on nesting success of Common Loons. *Journal of Wildlife Management* 77 (8): 1626–38.

Epilogue

We were assembled on a small lake in Seney National Wildlife Refuge, and it was warm, humid, and buggy, a true test of fortitude and concentration. We were there to catch a loon. For the past ten days this group of Earthwatch volunteers and I had worked together to catch and "work up" several loons—banding, taking blood and feather samples, weighing, data recording. A couple of weeks prior, I had come up with the idea of catching, working up, and releasing a loon without anyone uttering a single word, not an easy accomplishment. How would we coordinate getting the loon out of the water? Who would hold it? Who would it carry it to shore? How would we communicate the weight and measurements of the bird? I believed it was possible if you could find the right group of people. I liked the challenge of pulling this off. But I had a deeper purpose: I wanted to show respect for this extraordinary and amazing bird we call the loon.

This was indeed the right group of volunteers, and we rehearsed our roles several times before we launched the boat from shore. The only audible sound was my net entering the water. I scooped up the loon and placed the net in the bottom of the boat. Robb put a towel over its head and held it in place until I was able to assist. I lifted the loon a few inches off the deck of the boat while Karen separated its feet from the net, freeing it entirely. I then lifted the loon completely off the boat and placed it on Mary's lap. She squeezed the wings against its body, and I positioned the head under her armpit, making sure to keep the bird's dangerous bill away from her. Satisfied that all was well, I drove the boat to shore. On shore, Mary and I sat in the middle of a tarp. We had been doing this for the past few nights, and each volunteer had a job. Samantha recorded the location and the name of the loon territory on the data sheet. This loon had been previously banded, so I showed her the metal band, and she wrote down the nine-digit number, plus the color band combination of each leg. I took the bill measurements with calipers, and Samantha wrote down the values on the data sheet. I looked at the data sheet and gave her a thumbs up. Neither of us spoke.

Next, I inserted a needle into the leg and drew blood, filling the vacutainer. Karen placed a cotton swab over the injection site. Then, with help from Karen and Robb, I repositioned the loon, opened the right wing slightly, separated the individual feathers, and clipped one at the base. I did the same for the left wing. Karen took the wing feathers and placed them in an envelope marked with the band number. Then I took the loon from Mary, lowered it into a sack, and lifted it off the ground and onto a scale. Samantha recorded the weight of the loon. I glanced over the data sheet one last time, confirming the data were entered correctly.

It was time to release the loon. I gently received it from Mary, walked to the lake's edge, and stepped into the water. I lowered the loon until its belly touched the water's surface, and then I relaxed my grip. We had had the loon for twenty-two minutes, close to an average time, but we had done our work in complete silence. We had pulled it off. I expected the loon to rush out of my hands and explode across the water, as they normally do, but not this loon, not this time. Instead, it sat calmly on the water, and after a few seconds, it wailed once, then slowly and serenely swam away.

For those of us who love and admire loons, seeing or hearing one makes for a great day, but I am in awe when I hold and touch one in the wild. That feeling never leaves me. The lessons these magnificent birds have taught me are many. And we still have much we can learn from them.

Acknowledgments

I owe a large thank you to Earthwatch Institute for financial support. In total, Earthwatch funded eleven years of study, fielding fifty-two research teams and 248 volunteer scientists. I could not have achieved as much had it not been for its belief in my research projects. And yet, I may have gotten the better part of the partnership because my life was so greatly enriched by not only interacting with the wonderfully talented staff and being part of such a great organization, but by having the pleasure to train, teach, and interact with Earthwatch volunteers. These volunteers blew me away with their dedication, knowledge, curiosity, and enthusiasm, and at their core they were amazing people who sacrificed and worked hard to gather data. They endured blackflies, horseflies, deerflies, ticks, temperature extremes (heat and humidity, cold and fog), sleep deprivation, and so much more. I wish I could list all their names, for so many were super helpful and amazing people, but I have space to list only a few: Sherry Abst, Michelle Cooke, Bonnie and Dick Mabee, and Claudia Seldon. Thank you.

Over the years I have had the distinct pleasure to learn and work beside remarkable biologists, some for only a day or two, and some for weeks, months, and even years. I wish to thank and recognize this latter group: Evan Adams, Aleya Brinkman, Allie Byrd, Mike Chickering, Alex Dalton, Ariel Davila, Chris DeSorbo, Andrew East, Rick Espie, Scott Ford, Jennifer Fox, Holly Gomez, Ginger Gumm, Bill Hanson, Eric Hanson, Peg Hart, Franske Hoekeme, Joe Kaplan, Kevin Kenow, Michelle Kneeland, Kristin Kovach, Ronald de Lange, Melissa Lockman, Darwin Long IV, Zack Maddox, Jay Mager, Damon McCormick, Cory Counard McNulty, Jason Mobley, Kevin Morlock, Larry Neel, Chris Persico, Mark Pokras, Daniel Poleschook, Pete Reaman, Lucas Savoy, Nina Schoch, Diana Solovyeva, Vince Spagnuolo, Nicole Stacy, Virginia Stout, Keren Tischler, Adam Turpen, Hannah Uher-Koch, Lucy Vlietstra, Brooks Wade, Charles Walcott, Steve Wilkie, Jeff Wilson, Mark Wiranowski, Ken Wright, and Dave Yates. Thanks also to my non-biologist colleagues, from whom I learned much: Todd Seither, Adam Tutt, and Jim Waytek. To anyone I omitted, my sincerest apologies.

To my undergraduate and graduate students, who have kept me inspired, thank you for the enthusiasm and unbridled spirit: Brandon Braden, Priscilla Carnaroli, Adrienne Dolley, Andrew East, Natalie Juris, Igor Malenko, Matt Schultz, and Hannah Uher-Koch. In particular, I recognize and thank Hannah for being an outstanding lead assistant for four years in the Gulf of Mexico. Without her leadership and communication skills, in addition to her competence in the field and unbridled enthusiasm, the research project would not have been as successful.

To my fellow loon researchers, your curiosity, dedication, and hard work resulted in numerous publications and communications that increased my knowledge and understanding of these birds, and the book is more complete because of your efforts: Walter Piper, Kevin Kenow, Jay Mager, Dave Evers, Charlie Walcott, Mark Pokras, Michael Meyer, Nina Schoch, Lee Attix, Jeff Wilson, Paul Spitzer, Terry Dalton, and Glenna Clifton.

I thank Daniel Poleschook, Ginger Gumm, and Darwin Long IV for their photographs, which greatly enhance *Loon Lessons,* but, more important, I thank them for their friendship. The following people also contributed photographs, for which I am appreciative: Roberta Olenick, Ryan Askren, Jari Peltomäki, Jeff Bucklew, Jonathan Fiely, Michelle Kneeland, and Mark Pokras. Mark Burton generated and produced the maps.

I owe a special thank you to Mark Grover for reading the entire manuscript (twice) and giving me many helpful edits, comments, and suggestions. He made them in a timely fashion, which helped the project move forward, and his keen insight about the natural world and attention to detail were greatly appreciated. Partnering with him led to a bonus, a rekindling of our friendship, for which I am eternally grateful. Sue Leaf read the entire manuscript and offered many useful suggestions to improve it. Both Nina Schoch and Kristin Kovach read earlier drafts of several chapters, and their comments and edits significantly improved them. The following colleagues, friends, or family members read parts of the manuscript, and I thank them for their time and helpful comments: Glenna Clifton, Dave Evers, Tiffany Grade, Carrol Henderson, Kevin Kenow, Darwin Long IV, Jay Mager, Olivia Paruk, Mark Pokras, Dan Poleschook, Stefania Strzalkowska, Kate Taylor, Charles Trost, and Steve Windels. Any mistakes that remain are entirely mine.

Three individuals stand out, and I would like to acknowledge each for their unique contribution to my research and development as a loon biologist. Jay Mager has been supportive of my research and efforts since 1993. He is a deep thinker, passionate about both loons and science, and despite being busy with his own research projects, he was always there if I needed a helpful ear. His modeling of humility and compassion might be what I remember most. In 2013 I met Brooks Wade, a naturalist from South Carolina, who blew me away with his enthusiasm for loons. I did not expect to meet a kindred spirit, especially so late in life, but because of a phone call, our paths crossed, and I am better for it. So, dear friend, thank you for your friendship but especially for invigorating my research. Finally, Dave Evers, with whom I have traveled a long and winding road beginning as freshmen at Lake Superior State University in 1980. I owe you more than thanks, a handshake, or a shout-out of gratitude for, frankly, you enriched my life and opened the door to my becoming a loon biologist. By including me in some of your research projects, you sculpted me and gave me experiences for which I am profoundly grateful. Thank you for everything.

My editor at the University of Minnesota Press, Kristian Tvedten, was enthusiastic and supportive of *Loon Lessons* from the beginning. I cannot thank him enough. I appreciated his approach in guiding me to clarifying my thoughts and still retaining my uniqueness as an author and storyteller. He was the consummate professional every step of the way, and I was extremely fortunate to work with him. I owe a debt of gratitude to the staff at the University of Minnesota Press who assisted in the preparation of the book.

My lifelong friend Duane Starzyk gave me his unwavering support and timely conversations, even if some occurred after I was exhausted from an all-night banding excursion.

I thank my parents, Phyllis and Walter, for their support and encouragement to follow my passions, even though my path as a biologist was unconventional and less familiar to them. And I thank my family, starting with my wife, Stefania Strzalkowska, for doing all the little things, which allowed me to focus and concentrate on this book, and my daughters, Olivia and Emily, for being super supportive of my fieldwork. It made my frequent departures (and returns) easier.

Resources for Loon Conservation

Minnesota Loon Monitoring Program
http://www.dnr.state.mn.us

LoonWatch, Sigurd Olson Environmental Institute
http://www.northland.edu/centers/soei/

Maine Loon Project, Maine Audubon
http://www.maineaudubon.org/

LoonWatch, Vermont Center for Ecostudies
http://vtecostudies.org

Montana Loon Society
http://montanaloons.org

Loon Preservation Committee
https://loon.org

Adirondack Center for Loon Conservation
https://www.adkloon.org

Minnesota Pollution Control Agency
https://www.pca.state.mn.us/water/citizen-water-monitoring

Biodiversity Research Institute
http://www.briloon.org/loons

National Loon Center
https://www.nationallooncenter.org

Index

Alberta, 59, 138, 201
anatomy, 4, 31-32, 37, 54
Arctic Loon, 7-10, 91-92
artificial nesting platforms, 178
aspergillosis, 162-63
Audubon, John James, ix, x, xiii
Audubon Society, 146, 194, 201, 206

Bald Eagle, 21, 32, 42, 67, 86, 90, 105, 120, 130, 147, 164, 168, 174
banding, of loons, 24, 42, 168, 211
behavior: cooperation, 75-76, 103, 115, 117, 119; interspecific, 10, 62; intraspecific, 10, 22, 51, 91; nocturnal, 90, 123; social, 117, 160, 165
blackflies, 77
botulism, 140-42
British Columbia, 138

California, xii, 7, 127, 147, 152-53, 158
calls, 78, 82-87, 90-91, 93, 98. *See also specific calls*
chicks, 55, 91, 99-102, 130, 139, 198-200; development, 101; plumage, 100
climate change, 41, 142-43, 200-203
clutch size, 70-72
Colymboides, 7
communication: long-distance, 22-23, 27, 85; short-distance, 91. *See* calls
Connecticut, 189, 201, 204
conservation, xiii, 154, 167, 174, 185, 195, 197, 202, 204-6, 217
cormorant, 6, 12-13, 18-19, 21-23, 38-39, 42-44, 55, 61, 75, 89, 93, 161
courtship, 49, 63, 64, 87, 91

Darwin, Charles, xii, 19, 203
Darwinian selection, 60, 133
DDT, 168
Deepwater Horizon oil spill, 145, 153, 176, 178-81, 185
diet, 11, 13, 20, 42, 43, 106, 158, 160, 164
disease, 163, 168, 202, 204
distribution, 8, 12, 57, 138, 146, 149, 151, 161, 165
diving birds, 3-4, 6, 19-20, 24, 32-33, 35-36, 39, 41-42, 44-45

Earthwatch, 51, 73, 81, 97-98, 103, 107, 113, 116, 123, 145, 151, 159, 176, 181, 183, 202, 211
eggs, of loons, 10, 41-42, 67-68, 71-77, 100, 140, 169-70, 191, 194-95, 206

feathers, 2-3, 8, 19-20, 24-25, 40, 42, 64, 76, 100-101, 120-21, 123, 156-57, 162, 167-68, 176-77
fidelity, 53, 152-154, 165
food: of adults, 106; of adults, during winter, 106, 158; of chicks, 104
foot waggle, 120-22, 156
fossils, 4, 6-7, 204

Gavia: *adamsii,* 7; *arctica,* 7; *immer,* 7; *pacifica,* 7; *stellata,* 7
global warming. *See* climate change
goby, round, 141-42
grebes, 3, 6, 11-13, 28, 36, 41, 45, 61, 63, 101, 120, 128, 130, 141, 147, 161
gular fluttering, 41-42

habitat, 13, 43, 51, 115, 118, 134, 178, 186, 189-90, 192-94, 197, 201, 203; freshwater habitat selection, 151-52; marine habitat selection, 146-51; marine vs. freshwater habitats, 164-65; selection, 57-60

Hesperornis, 4–5
hoot, 91–92, 161
human recreation, effect on loons, 164, 192–93, 198
hybridization, 11

incubation, xi, 53, 67, 73–74, 88, 122, 191, 201; length, 75–76; patterns of next exchange, 76–78
islands, as habitat, 10, 21, 57, 69

juveniles, 28

lead: in fishing tackle, 171, 175; legislation, 173, 174; poisoning, 171–72
Loon Preservation Committee, 195, 206, 217

Maine, xii, 54, 67, 130–31, 140, 146, 153, 155, 157–60, 162, 163, 173, 178, 189, 191, 193–94, 200, 202, 204
malaria, 202, 204
Manitoba, 135–37
Massachusetts, 54, 147, 148, 173, 176, 200–201, 204
mercury: in environment, 168–70; legislation, 170, 174, 182; in loons, 2, 75–76, 164, 167, 174, 182, 203, 205
Michigan, xi, 24, 31, 49, 61
migration, 127–40
Minnesota, 58, 89, 130–32, 135, 137, 174, 178, 185, 189–90, 194–95, 198, 201, 202
Minnesota Department of Natural Resources, 69, 198, 203, 206
molting, 156–57
mussels, 141–42

natural selection, 19–20, 29, 32, 36–37, 53, 60, 68–69, 71–72, 85, 87, 99, 135, 137, 139, 155
nest: failure, 69, 177, 70, 173, 190, 196–97, 201; location, x, 21, 43, 68–69, 201; selection, 68–70
New Hampshire, 135, 167–68, 170, 173–74, 194–95, 202, 204, 217
New York, 173, 193–94, 200, 202, 217
niche, 9, 11–13, 45, 51
nonbreeders, 115, 117
Nova Scotia, 131
nursery, 193

oil gland, 24–25, 77
oil spills, 10, 43, 145, 153, 177–80, 186; effect on loons, 176–78, 183, 184–85
Ontario, 58, 101, 137, 191, 202

Pacific Loon, 1, 7–10, 45
parasites, 162–63, 202
parental care, 53, 102–3, 109, 199
penguins, 19, 34, 44–45, 76, 101
plumage, ix, 3, 7, 11, 18, 20–24, 28, 75, 139, 177, 180; chicks, 100–101
population, 11, 77, 146, 149, 156, 164, 170, 190, 203–6
predators, of loon, 2, 21, 24, 42, 69–70, 155, 194
productivity, 164, 190, 193–94, 197, 202

raccoons, 21, 69–70, 194
Red-throated Loon, 1, 7–11, 13, 45, 91–92, 148
restoration, 185, 189, 197

salt gland, 43
Saskatchewan, 59, 129, 136–39
satellite transmitter, 54, 127–29, 137, 153, 200
Seney National Wildlife Refuge, xi, 31, 49, 51, 68, 81, 88, 97, 113, 115–18, 155, 173, 205, 211
shoreline development, 190, 192, 194–95, 198, 204
sibling rivalry, 98–99
social gathering, 115–21

territories, 59–61, 76–78, 86–90, 105, 107, 109–10, 115–16, 118, 120–21, 124; defending, 61–63, 93, 104, 114, 119; types of, 58, 74, 76
trade-offs, 33, 67, 71, 78

tremolo, 81, 85, 87, 90–93, 113–14, 116–17, 196

Vermont, 173, 206, 217

wail, 3, 49, 81, 85–87, 90–93, 105, 113, 212
Washington, 138, 147, 170, 172–73, 205
West Nile virus, 202
winter range, 9, 56, 131, 158, 201
Wisconsin, 42, 55, 63, 67–69, 102, 107, 115, 130, 132, 137, 168–70, 178, 189, 204, 206

Yellow-billed Loon, 1–3, 7–11, 13, 91–92
yodel, 3, 8, 81–82, 85, 87–93, 97, 113, 116–17

James D. Paruk is one of the world's leading experts on the Common Loon. He has studied breeding loons and wintering loons in several states and in Canada for almost thirty years. After the Deepwater Horizon oil spill in 2010, he studied the health and recovery of wintering loons in Louisiana for Earthwatch Institute. He has served as vice president of the North American Loon Fund and chair of the research committee for LoonWatch, the flagship program of the Sigurd Olson Environmental Institute at Northland College in Wisconsin. He is professor of biology at St. Joseph's College, adjunct professor of biology at the University of Southern Maine, and adjunct senior research scientist at the Biodiversity Research Institute in Portland, Maine.